Objective Resilience

Other Titles of Interest

Objective Resilience: Policies and Strategies, **MOP 146**, edited by Mohammed M. Ettouney, Ph.D., P.E. (ASCE, 2021). Examines policies and strategies related to community and asset resilience and provides civil infrastructure stakeholders with a comprehensive, recommended set of practices. (ISBN 978-0-7844-1588-7)

Objective Resilience: Objective Processes, **MOP 147**, edited by Mohammed M. Ettouney, Ph.D., P.E. (ASCE, 2021). Illustrates some of the objective processes that are used to manage community and asset resilience and provides infrastructure stakeholders with a comprehensive set of practices. (ISBN 978-0-7844-1589-4)

Objective Resilience: Applications, **MOP 149**, edited by Mohammed M. Ettouney, Ph.D., P.E. (ASCE, 2021). Provides different applications that aim to enhance community and asset resilience from the community viewpoint. (ISBN 978-0-7844-1591-7)

ASCE Manuals and Reports on Engineering Practice No. 148

Objective Resilience

Technology

Sponsored by the
Objective Resilience Committee of the
Engineering Mechanics Institute of the
American Society of Civil Engineers

Edited by
Mohammed M. Ettouney, Ph.D., P.E.

Published by the American Society of Civil Engineers

Library of Congress Cataloging-in-Publication Data

Names: Engineering Mechanics Institute. Objective Resilience Committee, author. | Ettouney, Mohammed, editor.
Title: Objective resilience. Technology / sponsored by the Objective Resilience Committee of the Engineering Mechanics Institute of the American Society of Civil Engineers; edited by Mohammed M. Ettouney, Ph.D., P.E.
Description: Reston, Virginia : American Society of Civil Engineers, [2022] | Series: ASCE manuals and reports on engineering practice; no. 148 | Part of a four book committee report comprised of: Policies and strategies; Objective processes; Applications; Technology. | Includes bibliographical references and index. | Summary: "MOP 148 examines the use of different technologies to enhance community and asset resilience and provides a comprehensive set of practices for infrastructure stakeholders"-- Provided by publisher.
Identifiers: LCCN 2021051381 | ISBN 9780784415900 (print) | ISBN 9780784483763 (PDF)
Subjects: LCSH: Reliability (Engineering). | Civil engineering--Standards. | Infrastructure (Economics) | Natural disasters--Risk assessment. | Emergency management--Government policy. | Organizational resilience.
Classification: LCC TA169 .E545 2022d | DDC 620/.00452--dc23/eng/20211118
LC record available at https://lccn.loc.gov/2021051381

Published by American Society of Civil Engineers
1801 Alexander Bell Drive
Reston, Virginia 20191-4382
www.asce.org/bookstore | ascelibrary.org

***Errata:* Errata, if any, can be found at https://doi.org/10.1061/9780784415900.**

ISBN 978-0-7844-1590-0 (print)
ISBN 978-0-7844-8376-3 (PDF)
ISBN 978-0-7844-8412-8 (ePub)

Manufactured in the United States of America.
27 26 25 24 23 22 1 2 3 4 5

MANUALS AND REPORTS ON ENGINEERING PRACTICE

(As developed by the ASCE Technical Procedures Committee, July 1930, and revised March 1935, February 1962, and April 1982)

A manual or report in this series consists of an orderly presentation of facts on a particular subject, supplemented by an analysis of limitations and applications of these facts. It contains information useful to the average engineer in his or her everyday work, rather than findings that may be useful only occasionally or rarely. It is not in any sense a "standard," however, nor is it so elementary or so conclusive as to provide a "rule of thumb" for nonengineers.

Furthermore, material in this series, in distinction from a paper (which expresses only one person's observations or opinions), is the work of a committee or group selected to assemble and express information on a specific topic. As often as practicable, the committee is under the direction of one or more of the Technical Divisions and Councils, and the product evolved has been subjected to review by the Executive Committee of the Division or Council. As a step in the process of this review, proposed manuscripts are often brought before the members of the Technical Divisions and Councils for comment, which may serve as the basis for improvement. When published, each manual shows the names of the committees by which it was compiled and indicates clearly the several processes through which it has passed in review, so that its merit may be definitely understood.

In February 1962 (and revised in April 1982), the Board of Direction voted to establish a series titled "Manuals and Reports on Engineering Practice" to include the manuals published and authorized to date, future Manuals of Professional Practice, and Reports on Engineering Practice. All such manual or report material of the Society would have been refereed in a manner approved by the Board Committee on Publications and would be bound, with applicable discussion, in books similar to past manuals. Numbering would be consecutive and would be a continuation of present manual numbers. In some cases of joint committee reports, bypassing of journal publications may be authorized.

A list of available Manuals of Practice can be found at http://www.asce.org/bookstore.

DEDICATION

This objective resilience manual of practice is dedicated to the essential workers who are exposed daily to the dangers of the COVID-19 pandemic. Included among the many groups of workers are the following: healthcare personnel, first responders, public safety officers, correction facility workers, food and agriculture, grocery store workers, teachers, US postal service workers, public transit workers, and many more people who work tirelessly to maintain a sense of normalcy in these unprecedented times.

CONTENTS

BLUE RIBBON PANEL (In Alphabetical Order)

Joseph Brennan, R.A. (New York), AIA, is an architect and digital practice evangelist who has worked on projects in various capacities for SHoP Architects, Populous, and Gensler. In addition, he is currently an adjunct assistant professor at Columbia University's Graduate School of Architecture, Planning, and Preservation and has taught design technology–focused courses at various institutions in New York. In addition to practice and teaching, Joseph has mentored students at Columbia University, as well as emerging businesses through the New Museum's New Inc. Incubator program.

James Brunetti, P.E., is currently director of operations for Absolute Civil Engineering Solutions in Ft. Lauderdale, Florida. James has more than 35 years of experience in structural mechanics, structural dynamics, design of structures for extreme loadings, fracture mechanics analysis, forensic engineering, and structural failure analysis. He holds a professional degree in engineering mechanics and an M.Sc. degree in structural engineering from Columbia University, as well as a B.Sc. degree in civil engineering from the University of Virginia.

Albert DiBernardo, P.E., ACC, is a consulting engineer. He is the past president of TAMS Consulting, Inc. and the past executive vice president/principal of Thornton Tomasetti, Inc. DiBernardo began his A/E/C career in 1974 and for more than 43 years worked on civil infrastructure projects worldwide, including water resources, airports, bridges, buildings, port facilities, and environmental projects. In the mid-1990s, he began teaching engineering professionals leadership and management and until 2016 served as an adjunct professor in the NYU Tandon School of Engineering graduate program for architects, engineers, and construction professionals. Today, DiBernardo is still serving professionals in the field as a certified life/career and business coach, business advisor, meeting facilitator, mentor, and teacher.

Ketan Dodhia, P.E., specializes in man-made hazards, blast and progressive collapse at Stone Security Engineering. For nearly 20 years as a structural engineer, Ketan has focused on protective design projects and worked on numerous sensitive public and private sector projects. In addition to working all over the World Trade Center sites, including the National September 11th Memorial and Museum, Calatrava PATH Hub Station, and WTC Towers 2, 3, and 4, Ketan has worked on the New York Stock Exchange, the new Smithsonian National Museum of African American History and Culture, and various US embassies worldwide. As a senior project manager at Stone, Ketan is responsible for project management and leadership of the structural design team on blast-resistant design assignments.

Christopher Doyle is a consultant on homeland security matters, an industry in which he has been immersed for 30 years. Doyle led disaster response and recovery operations both in the field and at the headquarters while with the Federal Emergency Management Agency (FEMA). With the creation of the Department of Homeland Security, Doyle went on to help in the stand-up of the Science and Technology Directorate, directing a portfolio of research and development focused on response and recovery, as well as infrastructure protection. In this role, Doyle led several initiatives to promote the notion of resilience by integrating protection from natural and man-made hazards throughout the built environment, with particular emphasis on critical infrastructure. He was presented with the Institute Award from the National Institute of Building Sciences for his leadership in this area.

Henry Green, Hon. AIA, has held several leadership positions in the building community, including serving as executive director of the Bureau of Construction Codes in the Michigan Department of Labor. Henry was a member of the Building Officials and Code Administrators (BOCA) Board of Directors for 10 years, holding the position of president in 1997. Henry was a founding member of the International Code Council Board of Directors, completing a term as a president in 2006. He served as a member of the National Institute of Building Sciences Board of Directors for 8 years, completing a term as chair in 2003 and serving as a president for more than 10 years. Henry also has served on numerous committees for other building industry organizations and is the recipient of numerous awards. Henry was recognized by the United States House of Representatives for his work as "a tireless advocate for building safety and enforcement of codes."

Ahsan Kareem, Ph.D., P.E., Dist.M.ASCE, NAE, is the Robert M. Moran Professor of Engineering and director of the NatHaz Modeling Laboratory at the University of Notre Dame. His work focuses on probabilistic characterization of dynamic load effects owing to wind, waves, and earthquakes on tall buildings, long-span bridges, offshore structures, and other structures via analytical and computational methods and fundamental experiments at laboratory and full scale.

Sarbjeet Singh, Ph.D., P.E., LEED, is a licensed civil engineer with more than 20 years of industry and research experience, knowledge, and expertise in diverse areas of structural engineering/dynamics related to buildings, bridges, railroad, tunnels, and infrastructure. Currently, Sarbjeet is working as Principal Engineer at Metropolitan Transportation Authority (MTA) Construction and Development, New York. His past industry experience includes working with Weidlinger Associates Inc. and AECOM. Sarbjeet executed his postdoctorate research in structural control at the Virginia Polytechnic Institute and State University, USA. Sarbjeet is currently an active member of the ASCE Technical Committee of Infrastructure Systems with the Transportation & Development Institute and is a member of many professional societies, including ASCE, the Earthquake Engineering Research Institute, and the American Institute of Steel Construction.

AUTHORS
(In Alphabetical Order)

Hamid Adib, Ph.D., P.E., F.ASCE, F.SEI, has extensive executive-level experience and perspective with a focus on the design and construction in delivery of complex projects. Adib served on many ASCE, EMI, and SEI committees/capacities, including the SEI board of governors.

Marlon Aguero is a graduate student at the Department of Electrical and Computer Engineering at the University of New Mexico (UNM). Aguero received his M.Sc. degrees in civil engineering from the UNM and industrial engineering from the Technical University of Valencia in Spain.

Raimondo Betti, Ph.D., is a professor in the Department of Civil Engineering and Engineering Mechanics at Columbia University. He received his Ph.D. degree from the University of Southern California. His research areas are structural health monitoring and suspension bridges.

Scott Campbell, Ph.D., P.E., is a senior vice president for structures and codes for the National Ready Mixed Concrete Association. He has more than 30 years of experience in industry and academia, specializing in nonlinear behavior, extreme loading conditions, and vibration.

Charles Carter, Ph.D., P.E., S.E., is president of the American Institute of Steel Construction, where he has worked since 1991. He has been active throughout his career in the technical activities of many professional organizations related to design and construction.

ZhiQiang Chen, Ph.D., M.ASCE, is an associate professor at the University of Missouri—Kansas City. His research interests are in civil infrastructure resilience and application of remote sensing technologies. He secured his Ph.D. from the University of California, San Diego, in 2009.

John Cross, P.E., spent 18 years as a senior staff member with the American Institute of Steel Construction. Since his retirement, he has continued representing the structural steel industry on committees developing green standards and codes.

Ron Eguchi, M.ASCE, is president and CEO of ImageCat, Inc. Eguchi has more than 30 years of experience in risk analysis and reduction studies. He has a particular interest in applying remote sensing technologies to disaster risk management.

M. Ettouney, Ph.D., P.E., F.AEI, Dist.M.ASCE, has 52 years of consulting experience in many areas, including in very low-to-ultra-high-frequency dynamics and man-made and natural hazards risk and resilience management. Lately, he has been concentrating on the use of game, decision, graph, and probabilistic graph theories (including developing theoretical interactions betwwn these theories) in infrastructure health, progressive collapse, and climate change.

Katherine A. Flanigan is an assistant professor in the Department of Civil and Environmental Engineering at Carnegie Mellon University (CMU). She also holds a courtesy appointment in the Department of Electrical and Computer Engineering at CMU. She is an expert in intelligent infrastructure systems.

Margaret Glasscoe, Ph.D., is a research scientist at the Jet Propulsion Laboratory. She has been working in the field of disaster response and decision support for more than 10 years and has particular interest in developing tools using physical models and Earth observations.

Emad Hassan is a graduate research assistant. His research focuses on natural hazard risk assessment of community's social institutions while capturing the dynamic interdependencies between physical, social, and economic sectors.

Charles K. Huyck is a founding partner of ImageCat, where he directs a team of engineers and scientists developing software and novel methods to assess risk. He has more than 25 years of experience in integrating geospatial technologies into disaster simulation and modeling tools.

Bandana Kar, Ph.D., M.ASCE, is an R&D Staff at the Oak Ridge National Laboratory. Her research interests are in community and infrastructure resilience modeling and risk communication. She secured her Ph.D. from the University of South Carolina, Columbia, in 2008.

Jerome P. Lynch, Ph.D., M.ASCE, F.EMI, is the Vinik Dean of Engineering at Duke University. He was previously on the faculty of the University of Michigan and served as the department chair of Civil and Environmental Engineering from 2017 to 2021. He received his Ph.D. from Stanford University in 2002.

Hussam Mahmoud, Ph.D., is the George T. Abell Professor in Infrastructure. He works on assessing community resilience extreme events, quantifying infrastructure damage to extreme single and multiple hazards, and evaluating, maintaining, and repairing deteriorated infrastructure.

Fernando Moreu, Ph.D., P.E., is an assistant professor at the Department of Civil, Construction and Environmental Engineering (CCEE) at the University of New Mexico (UNM) at Albuquerque, New Mexico. He received his M.Sc. and Ph.D. degrees from the University of Illinois at Urbana-Champaign.

Ali Naghshineh is a civil and structural engineer and has many years' experience in both industry and academic fields. His main interests include project management, rehabilitation/retrofit of buildings and infrastructures, structural health monitoring, earthquake engineering, structural resilience, and dynamics.

Roya Nasimi is a Ph.D. candidate at the University of New Mexico. She received her M.Sc. degree in structural engineering. Her current research involves developing a noncontact monitoring system that can be mounted on a UAV for safer and more cost-efficient inspections.

Maryam Nazari, Ph.D., is an assistant professor in the Department of Civil Engineering, California State University, Los Angeles. Her research interests lie in the areas of earthquake engineering, sustainable civil infrastructures, and structural health monitoring and vibration control.

Tabitha Stine joined Nucor in 2020 leading their Construction Solution Services group. Previously, Tabitha was a vice president at AISC spending 16 years speaking on innovations in steel, construction trends, resilience, and sustainability to the AEC community.

Fariborz M. Tehrani, Ph.D. (UCLA'08; AUT'93; SUT'90), is professor and ESCSI director with research and practice experiences in sustainable and resilient structural engineering, mechanics, and materials. He is the recipient of ASCE and CHESC Awards, serving ASTM, ACI, and ISI.

PREFACE

Engineering is a balance between analysis and design. Objectivity forms, mostly, the basis of mathematics and science, which form, mostly, the basis of analysis. Subjectivity forms, mostly, the basis of art, intuition, and imagination, which form, mostly, the basis of design (see Figure 1). Achieving a proper balance between subjectivity and objectivity during

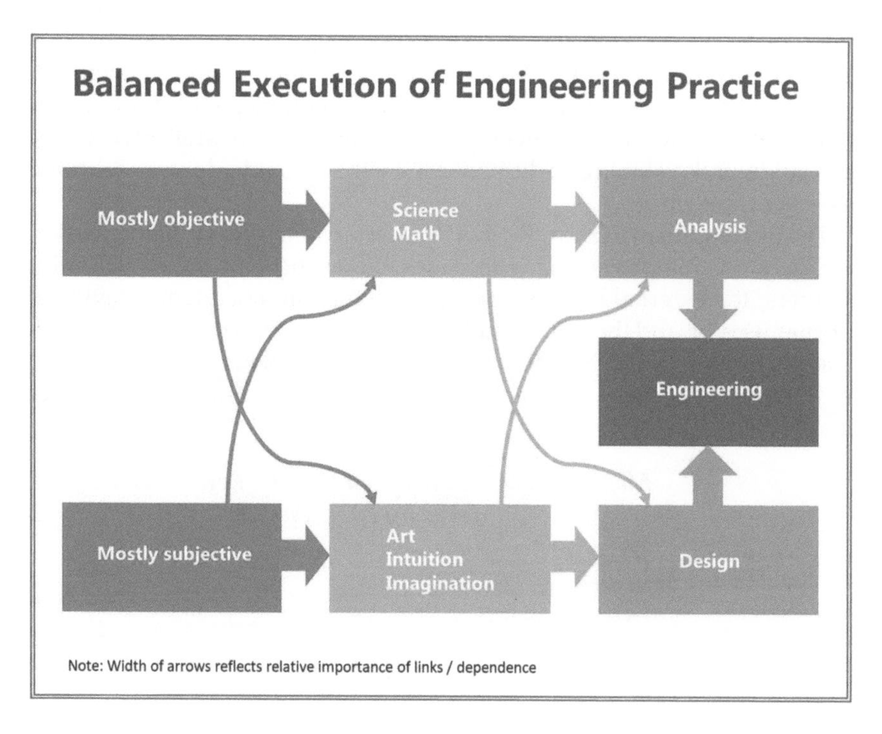

Figure 1. Balanced Execution of Engineering Practice.

the engineering process will ensure an optimal product. This is true especially for complex products that have multitudes of different types of components. Admittedly, community and asset resilience is a complex issue, and as such dealing with it from the engineering viewpoint will require a proper balance between objective and subjective processing.

The Objective Resilience Committee (ORC) of the Engineering Mechanics Institute (EMI) of ASCE was formed in 2015 to help achieve a balanced resilience treatment, especially from an objective viewpoint. Soon after its formation, the ORC initiated the development of an Objective Resilience Manual of Practice (OR-MOP) in 2016. The main objective of the OR-MOP is to provide a comprehensive basis of recommended practices that can help enhance community and asset resilience, while emphasizing the objective side of such practices. The developers of the OR-MOP quickly realized that because of the wide-ranging extent of community and asset resilience, the OR-MOP needed to split its focus into four basic categories: (1) Policies and Strategies, (2) Objective Processes, (3) Technology, and (4) Applications.

This book examines the use of different technologies to enhance community and asset resilience. It aims at providing a comprehensive set of practices, after presenting and discussing the basis for those practices. It is recognized that this OR-MOP is limited, given the limiting factors of space and time, especially in view of the aforementioned wide range extent of resilience. However, the developers hope that the OR-MOP can be used as a guide in developing additional MOPs that would address additional aspects of resilience.

The development of the OR-MOP took almost five years. Many worked tirelessly on this project. This includes the authors of the contributing chapters, the external blue ribbon panel, which independently reviewed the manuscript, and the ASCE Publications editors who provided valuable insights and feedback. Special thanks to Dr. Amar Chaker, the EMI director, for his efforts and help, without which this OR-MOP could not have been possible.

Mohammed M. Ettouney, Ph.D., P.E., F.AEI, Dist.M.ASCE
February 2021. West New York, New Jersey

INTRODUCTION

There are several popular definitions for resilience, including NIAC (2009), NSC (2011), or the Office of the Press Secretary (2013). For example, NIAC (2009) defined infrastructure resilience as follows:

> Infrastructure resilience is the ability to reduce the magnitude and/or duration of disruptive events. The effectiveness of a resilient infrastructure or enterprise depends upon its ability to anticipate, absorb, adapt to, and/or rapidly recover from a potentially disruptive event.

As defined, resilience represents a major issue for society, given the magnitude of disaster costs of different kinds. Recognizing the needs of society to build and sustain resilient assets and communities, stakeholders (for example federal, state, and local officials, business owners, professionals, educators, and researchers) devoted considerable effort, time, and expense examining asset and community resilience. Given the wide range of factors that affects resilience, knowledge gaps of the subject are still significant. Similar to most important topics, treatment, handling, and communicating resilience-related matters started with a subjective basis. Objective developments lagged their subjective counterparts; however, these developments have been gaining momentum in the past few years. One primary reason for the elevated interest in resilience-related objective processes is that without adequate objectivity, it will remain difficult to provide optimal policies and strategies that aim at delivering practical asset and community resilience at reasonable costs.

Recognizing the needs for comprehensive and practical objective views of asset and community resilience, the Objective Resilience Committee (ORC) of the Engineering Mechanics Institute (MEI) of the American Society of Civil Engineers (ASCE) embarked on developing an Objective

Resilience Manual of Practice (OR-MOP). The MOPs of ASCE aim at providing discussions, overviews, developments, and/or best practices concerning different topics. To better attain the stated goals, the OR-MOP endeavors to explore and discuss some of the many issues regarding objective resilience. The OR-MOP also strives to provide best practices sections in all the resilience-related subjects it covers. The *OR-MOP* attempts to address the intersection of three different areas: resilience (*Re*), civil infrastructure (*CI*), and objective processes (*OP*); see Figure 1. In a set-theory formalism, we can express *OR*-MOP as

$$OR - MOP \equiv Re \cap CI \cap OP \tag{1}$$

Because of the different nature of the chapters of the OR-MOP, we expect that the extent of their treatment of *OP* would vary.

To cast as wide a net for resilience-related objective issues as possible, which is not an easy task in itself, the OR-MOP is subdivided into four books. Each book will examine objective resilience from different viewpoints. Figure 2 illustrates the general subjects of the four books.

Given the essential role of technology in modern societies, we expect that technology will play a central role for providing resilient assets and communities. Because of this, we devote this book, as a part of the

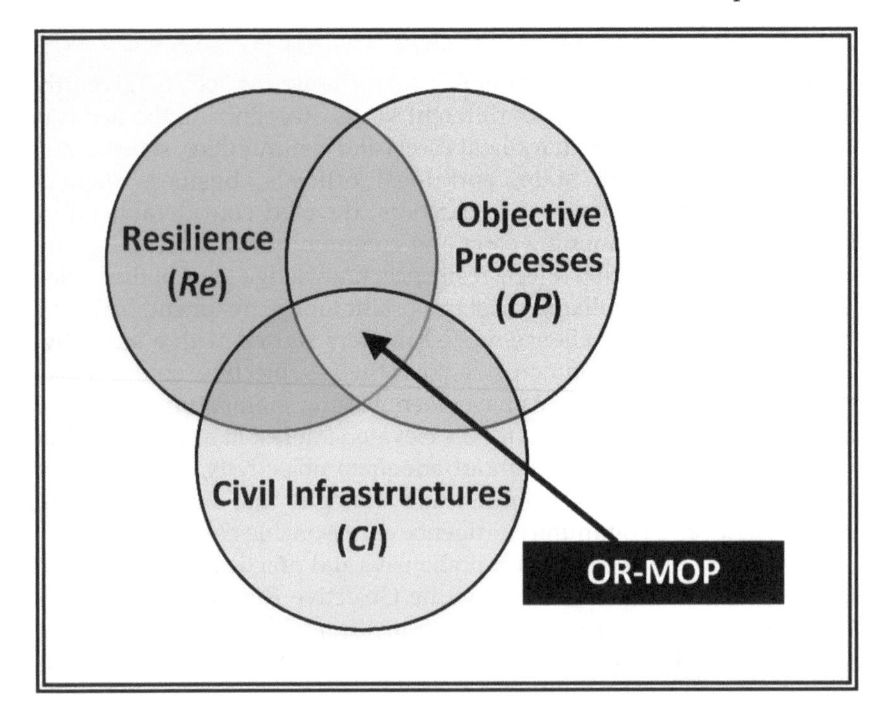

Figure 1. Confluence of domains of the OR-MOP.

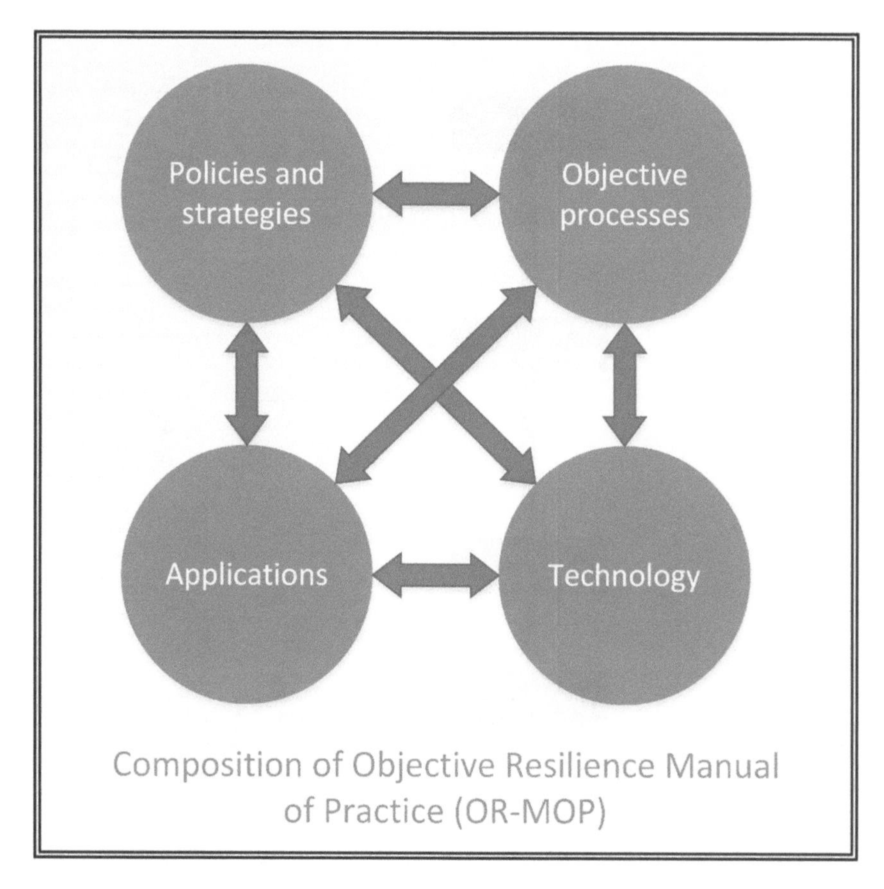

Figure 2. Composition of the OR-MOP.

OR-MOP, to the use of technology in achieving resilient assets and communities. The book looks at four segments of technology: materials (specifically construction materials), resilience monitoring (*ReMo*), hazards, and construction. Chapters 1 and 2 overview the role two main construction materials, concrete and steel, play in producing resilient assets and communities. Chapter 3 studies *ReMo* as applied to highway bridges, whereas Chapter 4 looks at the *ReMo* of railroad systems. Both chapters approach *ReMo* as a superset of the popular subject of SHM. Recognizing that resilience should always be related to a particular hazard (or a group of hazards), we devote Chapters 6 and 7 to the use of technology to enhance the resilience of assets and communities (as embodied in healthcare systems), respectively. Specifically, chapter 6 looks at enhancing asset resilience through improvements of its robustness. Chapter 7 studies enhancing community (healthcare systems) through response and recovery. The two chapters use earthquakes as the controlling hazard. The

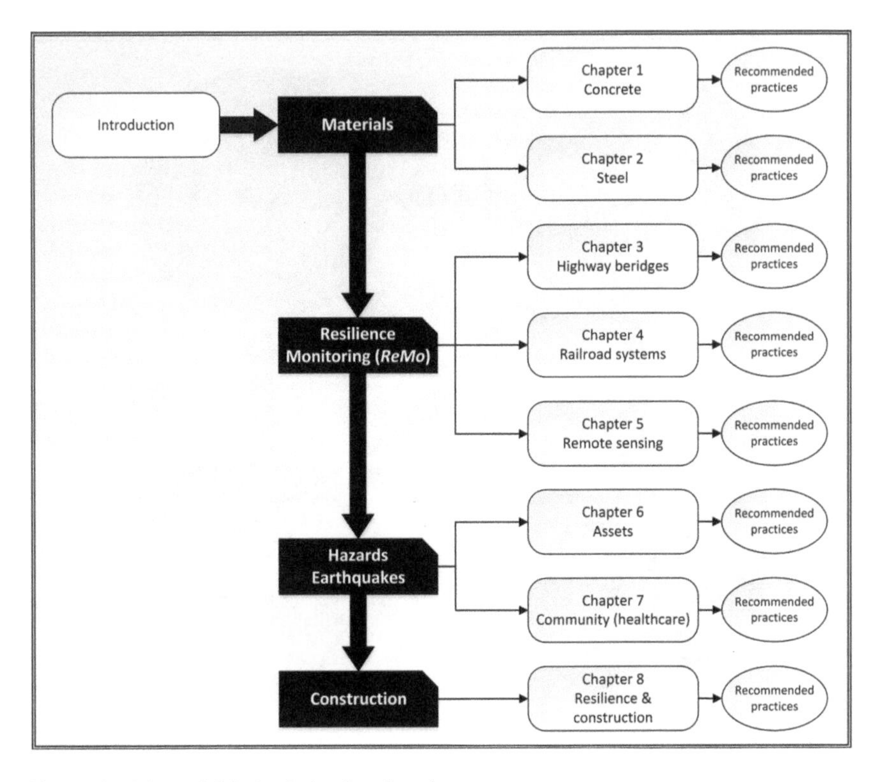

Figure 3. Map of this book (technology).

final chapter in this book overviews issues relating to resilience and construction. All chapters will propose a set of recommended practices at the conclusion of each chapter. See Figure 3 for a map of the organization of the book.

The intended readers of this OR-MOP include all civil infrastructure stakeholders, which may broadly include the following:

- Public and private civil infrastructures organizations (transportation, water resources, bridges, healthcare, and so on),
- City, county, and state officials,
- Emergency managers,
- Public safety personnel,
- Facility managers,
- Security consultants,
- Engineers, architects, and other design professionals,
- Educators, and
- Researchers.

Even though there are a wide range of objective complexities covered in the chapters, a deep knowledge of these objective topics is not required to achieve familiarity and benefit from the content. For readers who may not have the time to go in depth in each subject matter, it is suggested that they initially become familiar with the "recommended practices" at the end of each chapter. Each reader can then look at the chapter in depth to learn the reasonings/sources of these recommended practices.

Note that ASCE Manuals of Practice (MOPs) are developed by ASCE technical committees, such as the ORC, under the direction of an ASCE sponsor such as the Engineering Mechanics Institute (EMI). The distinguishing characteristic of an MOP, including this one, is that each one undergoes peer review by a Blue Ribbon Panel of experts before final approval is sought from the appropriate executive committee. Thus, the peer review by the Blue Ribbon Panel gives added weight to the MOP.

Mohammed M. Ettouney, Ph.D., P.E., F.AEI, Dist.M.ASCE

REFERENCES

NIAC (National Infrastructure Advisory Council). 2009. *Critical infrastructure resilience final report and recommendations*. Washington, DC: NIAC.

NSC (National Security Council). 2011. "Presidential Policy Directive/ PPD-8: National Preparedness." *Presedential Policy Directive*. Accessed May 26, 2018. https://www.dhs.gov/presidential-policy-directive-8-national-preparedness.

Office of the Press Secretary. 2013. "Presidential Policy Directive/PPD-21: Critical infrastructure security and resilience." *Presedential Policy Directive*. Accessed October 20, 2019. https://www.dhs.gov/sites/default/files/publications/PPD-21-Critical-Infrastructure-and-Resilience-508.pdf.

CHAPTER 1
RESILIENCE OF CONCRETE

Scott Campbell

1.1 INTRODUCTION

1.1.1 Definition

Resilience, in a statement adopted by many industry groups, is the ability to prepare and plan for, absorb, recover from, and more successfully adapt to adverse events. This seems very straightforward, but what does it mean in practice?

The ability to absorb adverse events is what is traditionally thought of as engineering design. That is, a hazard is identified, an appropriate design level is determined, and the project is designed to achieve a specified level of performance when subjected to the event. In general, the specified performance is prescriptive; deals only with life safety; and is required by standards, codes, and law. This traditional design process and performance measurement can cause problems when trying to achieve higher levels of resilience as attempts to move beyond life safety are met with resistance because "code compliant" is all that is required by law.

Preparing and planning for hazards is, at least partially, a component of the typical design process. The hazards and design levels specified in codes and standards are determined by deciding which hazards are present in a given location and what hazard magnitude should be resisted for the life safety design requirement. Typically, these hazard levels are based on historical data, but some efforts are currently underway to project into the future to determine design levels for the life of a project.

Recovery from a hazard is perhaps the most intriguing aspect of resilience. In general, it is not realistic to design for no damage under any magnitude of hazard. It is expected that there will be at least some damage when a project is subjected to low-probability hazard levels. Exactly how to set appropriate damage measures and determine how to measure the expected damage, are the subjects of ongoing debate and discussion. However, this is the heart of objective resilience—how resilient is your project?

Current thinking in many organizations is that "functional recovery," the time required for a specified level of functionality to be restored after an event, should be the measure of resilience. The amount of damage that can be sustained or how much repair must be performed to achieve a given level of functionality, is a multifaceted problem but is essentially technical in nature. The amount of functionality needed for a given project, however, is a political, economic, and sociological problem.

Successfully adapting to adverse events can take on many forms. One aspect is simply the good design practice of anticipating that building or infrastructure requirements may change over time and designing to allow for expansion or repurposing. One example is designing a structural building system to allow for the addition of extra stories without the need to strengthen the existing columns. The other aspect is to adapt to changing conditions, with particular emphasis currently being placed on climate change. This can include designing for greater frequency/magnitude of flood, high wind events, or sea-level rise.

1.1.2 Sustainability and Resilience

Sustainability is often confused, sometimes deliberately, with resilience in discussions of policy and design standards. Although sustainability is an important concern and plays a part in the resilient design, the terms are not interchangeable, and their goals are sometimes in conflict. One of the major components of sustainability, energy efficiency, can directly lead to increased resilience through reduced demands, which, in turn, allows for more cost-effective alternatives to the energy grid (e.g., generators) and also reduces demand on the infrastructure to provide sufficient energy to return to functionality.

However, it must be understood that not all sustainability goals lead directly to increased resilience. For example, windows are very inefficient in terms of thermal transmission and account for a large amount of the energy lost through the building envelope. So, to reduce energy use in a building, a designer would reduce the amount of glazing. However, doing so increases reliance on artificial lighting, which might render the building unusable if the power is out for an extended period.

Moreover, it must be recognized that for some structures, sustainability is, at best, a minor consideration. Examples could include hospitals, emergency response centers, military facilities, major bridges, and others. Although the design of these structures will consider sustainability concepts such as energy use, embodied carbon, and life-cycle environmental

Figure 1-1. Thermal efficiency of insulated concrete form (ICF) and wood-framed houses (foreground building and background building, respectively).
Source: Copyright Logix Brands.

costs, at the end of the day, the requirement that they remain functional after an adverse event will take precedence.

Concrete is ideally suited to meeting both resilience and sustainability goals. The potential savings in energy use are clearly evident in Figure 1-1, where a home built with insulating concrete forms and one that is traditionally framed and insulated have their heat loss measured, with a clear difference in thermal transmission. In addition, cement has made tremendous strides with respect to energy use during production, and when the true global warming potential from the cradle to the grave is considered, concrete performs similar to other building materials.

1.2 OBJECTIVE RESILIENCE

The preceding discussion illustrates that there are two primary ways to achieve true resilience in structures: (1) design to minimize damage during an event, and (2) design to allow for fast recovery from an event. In most cases, the resilient solution will be a combination of the two approaches. For most structures, it is not cost-effective, or even feasible, to design to have no damage under virtually any scenario. However, failing to provide an adequate level of resistance would also render any recovery efforts moot—there would be nothing left worth recovering.

Therefore, in any resilient design scenario, it is necessary to balance resistance and recovery in a way that makes sense for the particular structure being designed. One example of how to do this can be seen in seismic design. Structures are designed to limit damage during a seismic event to specific areas of the structure and to avoid significant damage in components where such damage may lead to greater damage throughout the structure. For a frame structure, specific plastic hinge zones are designed in the beams, whereas the columns are designed to remain essentially elastic. The idea is that failure of a column would be potentially catastrophic, whereas the beams may undergo significant nonlinear behavior without loss of their ability to carry gravity loads. The beams are also easier to repair, often without taking the entire structure out of service. By limiting the damage areas, the repair, and thus recovery time, is minimized. Taken together, proper design using these principles will lead to a more resilient structure.

Although the principles of objective resilience are clear and easy to understand, quantifying resilience is an area that needs significant advancement. The ability of the engineering community to predict damage during specific events is relatively well established, at least for certain hazards. However, predicting the recovery time and cost after an event is not nearly as well developed.

The remainder of this chapter discusses how concrete is ideally suited to addressing resilience for a variety of hazards, including examples and technical requirements. The resistance of structures to the hazards will be discussed, and a summary of repair/recovery will also be provided.

1.2.1 Hazards

The ability of concrete structures to resist hazards, both natural and human-induced, is explored in this section. Design aspects are briefly described along with the relevant standards, and examples of structures resisting the hazards are presented. The focus is primarily on buildings, but infrastructure examples will also be covered.

1.2.2 Earthquakes

Properly designed concrete structures are able to withstand even extremely strong seismic motions. This is reflected in the design standards. ASCE 7 (ASCE 2017a) allows *R*-values, which represent the ability of a structural system to undergo large nonlinear deformations without failure, up to 8 for concrete frames and 6 for concrete shear walls.

Obtaining adequate ductility in concrete members to obtain minimal damage during a seismic event consists primarily of providing confinement and connection detailing. For example, in high seismic zones, there are regions of a structural wall that are more likely to experience yielding of the longitudinal reinforcing (ACI 2019b). In those regions, it is not

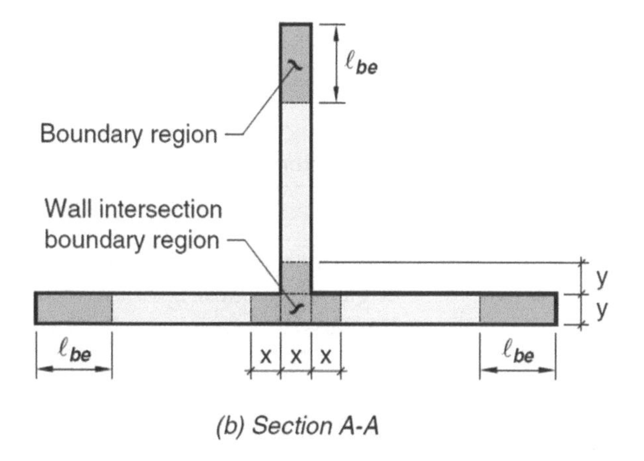

Figure 1-2. Wall zones where lap splices are not permitted in seismic regions. Source: ACI 318-19 (ACI 2019).

permitted to have lap splices (Figure 1-2) because it is unlikely that the required ductility could be developed in the splice region.

The importance of transverse reinforcement and the difference in performance obtained based on a design decision can be seen in Table 1-1 for reinforced concrete beams. The first takeaway is that there is a significant difference in the performance between beams with "conforming" and "nonconforming" transverse reinforcing. In this context, "conforming" means that within the expected plastic hinge zone of the beam, the transverse reinforcement is spaced less than the maximum allowable. For a given expected damage level, the allowable deformation before experiencing that damage increases substantially with conforming reinforcement. For example, the first and fifth lines of Table 1-1 have identical tension/compression reinforcement limits and shear requirements, yet the beams with the conforming transverse reinforcement can undergo double the end rotation (0.010 versus 0.005 rad) before experiencing minimal damage at the immediate occupancy damage level. Similar patterns in performance based on proper detailing are observed for concrete columns, joints, slabs, and walls, with transverse reinforcement and/or reinforcement continuity being essential if better performance is desired.

The result of using concrete for design in seismically active regions is that the damage owing to an earthquake can be reduced to the desired level and that the damage can be limited to areas of the structure that will not lead to collapse and that can be more easily repaired. The relative stiffness of concrete structures also tends to reduce the damage to nonstructural components such as interior walls, although that reduction is not well quantified. Given that nonstructural components make up a

Table 1-1. Reinforced Concrete Beam Modeling and Acceptance Criteria.

			Modeling parameters[a]			Acceptance criteria[a]		
			Plastic rotations angle (radians)		Residual strength ratio	Plastic rotation angle (radians)		
						Performance level		
Conditions			a	b	c	IO	LS	CP
Condition I. Beams controlled by flexure[b]								
$\frac{\rho-\rho'}{\rho_{bal}}$	Transverse reinforcement[c]	$\frac{V}{b_w d\sqrt{f'_c}}$ [d]						
≤0.0	C	≤3(0.25)	0.025	0.05	0.2	0.010	0.025	0.05
≤0.0	C	≥6(0.5)	0.02	0.04	0.2	0.005	0.02	0.04
≥0.5	C	≤3(0.25)	0.02	0.03	0.2	0.005	0.02	0.03
≥0.5	C	≥6(0.5)	0.015	0.02	0.2	0.005	0.015	0.02
≤0.0	NC	≤3(0.25)	0.02	0.03	0.2	0.005	0.02	0.03
≤0.0	NC	≥6(0.5)	0.01	0.015	0.2	0.0015	0.01	0.015
≥0.5	NC	≤3(0.25)	0.01	0.015	0.2	0.005	0.01	0.015
≥0.5	NC	≥6(0.5)	0.005	0.01	0.2	0.0015	0.005	0.01

[a]*Length of yield plateau (rad)*

[b]*Plastic deformation at failure (rad)*

[c]*Ratio of residual strength to yield strength*

Source: ASCE 41-17 (ASCE 2017b).

Notes: IO = Immediate Occupancy, LS = Life Safety, CP = Collapse Prevention.

large part of the cost of buildings, and their damage may render the building unusable, reduction of nonstructural damage will also lead to greater resilience. Limiting of damage, and therefore recovery time/cost, for both structural and nonstructural components means that proper design using concrete can objectively enhance the resilience of structures.

1.2.3 Hurricanes

Recent hurricane events have demonstrated the objective resilience of concrete construction during high wind events. For example, Figure 1-3 shows the aftermath of two hurricanes, where the only buildings left standing, with minimal damage, are those made primarily from concrete and/or masonry. Many of the structures that formerly stood in the locations depicted in Figure 1-3 were designed to modern building codes.

Figure 1-3. Damage from recent hurricanes: (a) Hurricane Katrina, (b) Hurricane Michael.
Source: (a) FEMA (2014), (b) US Department of Defense.

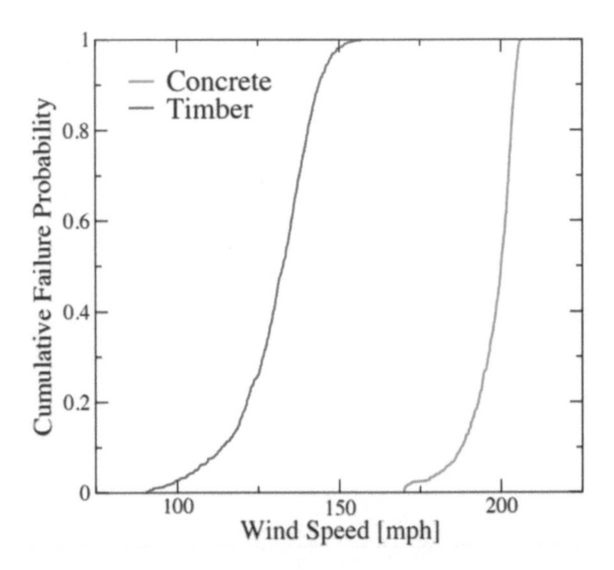

Figure 1-4. Collapse probability during a high wind event.
Source: Keremides et al. (2018).

The wind load provisions of these codes are intended to prevent the loss of life during an event but not necessarily to protect the property. Although there was very little loss of life in modern hurricanes in the United States, this is primarily because of the advance warning of an approaching hurricane and the subsequent evacuation of the areas expected to be the hardest hit. From Figure 1-3, it can be inferred that if these buildings were occupied at the time of the hurricane, it is certain that many more deaths and serious injuries would have occurred even with buildings designed to the latest building codes.

The additional resilience in concrete structures is owing primarily to having capacity well beyond the design strength. A recent study by the Massachusetts Institute of Technology (Keremides et al. 2018) demonstrates the better performance at ultimate strength for concrete structures. Figure 1-4 illustrates the collapse fragility curves for commercial wood-framed and concrete structures. Both buildings were designed to the same wind speed, but the performance is obviously quite different. At the design wind speed of 130 mi/h (58.1 m/s), the wood structure has a 55% probability of collapse, whereas the concrete structure collapse probability is essentially zero.

1.2.4 Tornadoes

Storm shelters have traditionally been constructed of concrete for a reason (Figure 1-5). Wind speeds in tornadoes can exceed those owing to even major hurricanes, causing significant, even virtually complete,

(a)

(b)

Figure 1-5. Examples of concrete structures surviving tornadoes with no/minimal damage: (a) home in Lafayette, Tennessee, survives a tornado that destroys surrounding homes; (b) concrete storm shelter survives a tornado with no damage. Source: FEMA (2014).

damage in the affected areas. Unlike hurricanes, there is very little advance warning for tornadoes, leaving people especially vulnerable if the structures they are occupying are not able to withstand the tornado wind forces.

In general, two items dominate the design of structures for resilience when subjected to tornadoes or other high wind events: wind speed and missile impact. Concrete structures are easily designed for high winds, even for wind speeds higher than specified in the building code. In addition, as seen previously, concrete walls have substantial additional capacity over wood structures even when both are properly designed for the same wind speed. Additional design concerns for concrete and masonry storm shelters are included in ICC 500, *Standard for Design and Construction of Storm Shelters* (ICC 2014). Two specific items are mentioned: (1) Requirements for additional protection at joints and other openings are required, except for "masonry control joints, masonry or concrete expansion joints or precast concrete panel joints 3/8 in. (9.5 mm) or less in width, sealed with joint material in accordance with TMS 602 for masonry or ASTM C920 for concrete." That is, typical concrete and masonry joints designed in accordance with industry standards do not require any additional protection. (2) Fire barriers and horizontal assemblies are required to have a minimum fire resistance rating of 2 h. Typical concrete walls and floors used in storm shelter design meet this requirement without additional materials.

Resistance to wind-borne debris impact is another area in which concrete construction typically meets the required criteria using standard designs. This is not true of most standard construction materials and designs, including those that meet minimum building code requirements (Texas Tech University 2004). The threshold missile speed at which damage from standard missiles might occur is listed in the test report cited and referenced by FEMA P-320 (FEMA 2014). The missile speed thresholds for typical concrete wall construction range from approximately 96 to 130 mi/h (42.9-58.1 m/s). Note that the missile design speed varies with wind speed but ranges from about 40% to 60% of the design wind speed for vertical assemblies (ICC 500) (ICC 2014).

1.2.5 Flood

Concrete has long been considered the gold standard material for flood control and protection. The combination of water resistance and mass makes it ideal for both control of water in infrastructure projects (dams, floodwalls, channels, and so on) (Figure 1-6) and protection of the interior of buildings and other facilities against the effects of flooding. Because of the inherent properties of concrete that contribute to flood protection, specific recommendations, such as those found in ASCE 24, *Flood Resistant*

(a)

(b)

(c)

Figure 1-6. Concrete flood control structures: (a) Gibraltar Dam, California; (b) concrete levee; (c) seawall in Galveston after the 1900 hurricane. Source: (a) Portland Cement Association, (b) Infrogmation/CC-BY-SA-3.0, (c) Library of Congress.

Design and Construction (ASCE 2014), are, in general, limited to suggestions regarding material choices. In addition to standard structural design, facilities designed in accordance with ACI 318-19 and ACI 350 (ACI 2006) to enhance resilience for flood events should

1. Choose the material properties of concrete and grout considering the potential for longer-term exposure to salt and other chemicals that are likely to be present during a flood event.
2. Specify a minimum 5,000 psi (34.5 MPa) concrete in areas within 3,000 ft (914 m) of the coastline to improve resistance to saltwater intrusion.
3. Consider the effect of loss of soil on foundations.

1.2.6 Blast

Concrete is ideal for blast-resistant design. The combination of strength, ductility, and mass means that concrete walls and frames are particularly suited for use in explosive environments. Design of concrete structures to resist blast loads is typically performed in accordance with ASCE 59-11 (ASCE 2011), Blast Protection of Buildings, or Unified Facilities Criteria documents as appropriate. An important point is that the blast design of a building utilizes a truncated form of performance-based design, in that the designer can choose performance levels (superficial to hazardous expected damage), and criteria are explicitly defined for each performance level. This process is related to the main tenets of resilience—directly to the concept of resisting the hazard and indirectly to reducing repair/return to functionality time.

The requirements for concrete for blast loads include limitations on ductility ratio or end rotation under the design loads, along with modeling and detailing requirements. The maximum ductility/end rotations (Table 1-2) are, in general, slightly higher than those for seismic response (Table 1-1) at comparable damage levels. Detailing requirements are similar to seismic detailing and are intended to allow for the development of adequate nonlinear behavior without premature failure. The detailing requirements include

- Use of only normal-weight concrete with a minimum strength of 3,000 psi (20.7 MPa),
- Restrictions on lap splices and additional transverse reinforcement at splice locations,
- Restrictions on the use of concrete shear strength for larger ductilities,
- Use of expected material properties rather than nominal for some calculations in ACI 318, and
- Larger minimum reinforcement ratios in some conditions.

Table 1-2. Maximum Response Limits for SDOF Analysis of Flexural Elements.

	Expected element damage							
	Superficial		Moderate		Heavy		Hazardous	
Element type	μ_{max}	θ_{max}	μ_{max}	θ_{max} (deg.)	μ_{max}	θ_{max} (deg.)	μ_{max}	θ_{max} (deg.)
Reinforced concrete								
Single-reinforced slab or beam	1	—	—	2	—	5	—	10
Double-reinforced slab or beam without shear reinforcement	1	—	—	2	—	5	—	10
Double-reinforced slab or beam with shear reinforcement	1	—	—	4	—	6	—	10

Source: ASCE 59-11 (ASCE 2011).

1.2.7 Fire

Fire protection engineering recognizes two basic differences in construction type—combustible and noncombustible. Concrete is, by definition, noncombustible, and therefore, inherently provides a higher level of protection against fires. This is reflected in the larger height and area limits in the International Building Code (IBC) (ICC 2018a) for noncombustible construction.

Damage to concrete structures from fire is typically minimal, and, although concrete can spall because of the differential expansion of the aggregate and mortar, major spalling of the concrete cover during a fire is typically because of expansion of the reinforcing at high temperatures. Although spalling can occur at temperatures as low as 350 °C, it more often occurs at much higher temperatures.

Some of the material factors that can increase the risk of spalling include high moisture content, high cement content, the use of silica fume, high compressive strength, and low permeability. These factors are all related to increased internal moisture or low permeability, both of which increase the risk of spalling through expansion of water in the voids and a lack of an escape path for water vapor. It is clear that if fire is a significant hazard for a structure, spalling risk can be reduced through the design of the concrete mix.

Reducing spalling owing to reinforcement expansion is a matter of reducing the temperature changes. The typical distribution of temperature within concrete is shown in Figure 1-7(a). It is clear that the temperature

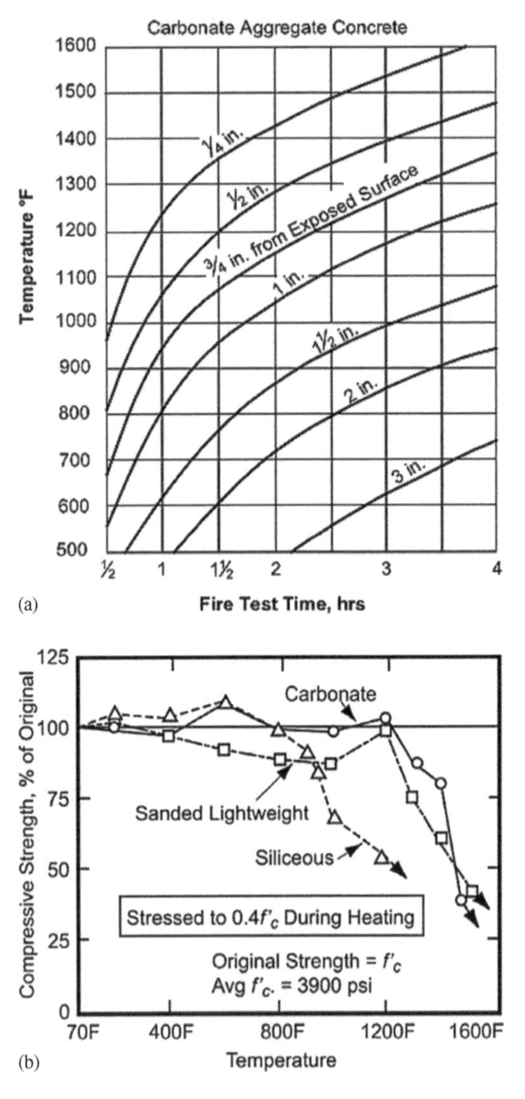

Figure 1-7. Effect of fire on concrete structures: (a) temperature within a concrete slab subjected to the standard fire, (b) variation of concrete strength with temperature.
Source: Abrams and Gustaferro (1968), Bilow and Kamara (2008).

decreases rapidly with depth, and, thus, providing adequate cover for the reinforcing is the basic design parameter for spalling prevention. Note that concrete strength decreases with increasing temperature [Figure 1-7(b)]. However, when the insulating properties of concrete are considered, it is obvious that only the outer portion of concrete members will suffer significant strength loss under any but the most severe and prolonged fire.

The discussion up to this point has been primarily aimed at structural fires. However, a growing concern for structures is wildfire. A substantial increase in damage owing to wildfires has been observed in recent years because of a combination of increased constructions in the wildland–urban interface and increased fuel available for fire growth (Figure 1-8). All the factors that make concrete structures safer during structural fires are also present in wildfires. Concrete structures are often the only things standing after a significant fire.

One factor that makes concrete even more resistant to excessive damage owing to fire is that even if the cover completely spalls off a concrete member, most of the strength and stiffness remains, making failure of the structural members from fire alone highly unlikely. In addition, repair of the concrete cover is not difficult and can often be performed while the structure is in service. Further discussion on the repair of concrete structures can be found later in this chapter.

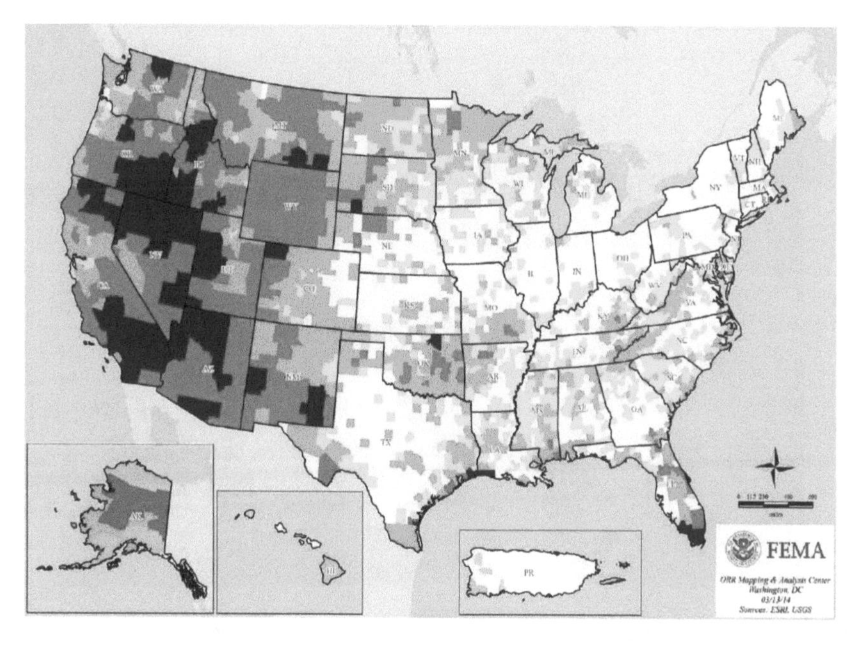

Figure 1-8. Wildfire activity by county, 1994–2013.
Source: FEMA (2014).

1.2.8 Other Hazards

The hazards listed in the preceding paragraphs are the most likely to be experienced by a majority of structures, but there are other hazards that concrete is also ideally suited to resisting. For example, vehicles do leave the roadway and impact buildings and bridges. The damage to concrete structures is significantly less than for other construction types, with multiple examples available in the press. An extremely unusual case is shown in Figure 1-9, where a rock fell and impacted a house. Although there is considerable damage, a major part of the house remains unaffected, which will lead to greater resilience because of the reduced time to return to full functionality. It is unlikely that the same event would have produced so little damage with less robust construction.

1.2.9 Recovery

The previous sections discussed designing concrete structures to resist natural and man-made hazards. Although it is possible to design concrete structures to resist almost any level of hazard with no or minimal damage, it is not necessarily practical or cost-effective to do so. The following sections discuss resources and techniques available to repair or rehabilitate structures, considerations for rebuilding, and issues surrounding decommissioning of structures.

Figure 1-9. Rock impact.
Source: National Park Service.

1.2.10 Repair/Rehabilitation

Multiple strategies are available to repair damaged concrete structures and to rehabilitate structures before an event to provide greater resistance or less anticipated damage and, hence, greater resilience. The most common strategies used depending on the nature and severity of damage or required strengthening include

1. Strengthening by bonding external steel or fiber-reinforced polymers to the structural elements,
2. Repair of cracks by grouting or the use of epoxy mortar or concrete, and
3. Repair of larger areas using epoxy concrete.

The International Existing Building Code (IEBC) (ICC 2018b) provides guidance on when repairs are needed and the requirements for those repairs. However, note that IECC is intended for preserving/restoring minimum life safety, and the provisions concentrate mainly on ensuring that the structure meets either the original or current building code requirements, depending on the extent and severity of damage.

Another resource for designing and implementing repairs to concrete structures is ACI 562, *Code Requirements for Assessment, Repair, and Rehabilitation of Existing Concrete Structures (ACI 562-19) and Commentary* (ACI 2019a). This standard is intended to be used in conjunction with the IEBC and IBC/IRC but has the advantage of being performance-based. That is, the owner or AHJ can specify any level of desired performance and then follow the outlined procedures to obtain a structure that meets those criteria. This is of particular importance where it is desirable for a structure to achieve a higher level of performance than the original design.

Extensive guidance is provided, often by reference to another standard, for the design of concrete repairs including

- Use of postinstalled anchors to transfer design forces (ACI 318),
- Guidelines on the size and shape of repairs (ICRI Guideline No. 310.1R) (ICRI 2008),
- Choice of expansion of joint materials (Parking Facility Maintenance Manual, (NPA 2015),
- Supplemental post-tensioning (ACI 318), and
- Fiber-reinforced polymer composites (ACI 440.1R and ACI 440.2R).

Durability of any repairs is crucial to their long-term success, and ACI 562 provides requirements for which conditions should be considered and what is considered to be unacceptable durability.

1.3 DECOMMISSION

When considering the resilience of structures, and in particular the sustainability aspect of resilience, what happens when the building is no longer needed must be considered. Concrete is often reused/recycled after demolition, and more than 140 million tons (1.27×10^{11} kg) of concrete are reused or recycled every year in the United States. Demolished concrete can be used in a variety of applications, including fill including roadway base, erosion control (riprap), gabions, and artificial reefs, and the aggregate can be reused in new concrete.

1.3.1 Adapt

An often overlooked aspect of resilience is the ability to adapt buildings to a new purpose, either on a temporary or on a permanent basis. Concrete is ideal for adaptation for a number of reasons. First, the relatively long spans in these buildings allow for reconfigurable interior spaces. This flexibility means that, for example, a building originally designed as an office or retail space can relatively easily be converted to other purposes such as housing. An excellent example is the conversion of retail space into health-care facilities. Concrete walls provide built-in benefits for health care such as excellent fire ratings and acoustic performance to meet privacy objectives.

Concrete structures also have guidance available for expansion. ACI 318 provides a standard for the design of anchorage including the development of new reinforcing. Although adding a story to a building is often limited by the capacity of the gravity system, this capacity can be increased using the techniques outlined in Section 1.2.10.

1.3.2 New Technologies and Advancements in Concrete and Masonry

Although concrete has been used since the time of the Roman Empire, the technology continues to advance in ways that increase the resilience of concrete structures. Some of these advances are discussed in more detail elsewhere in this chapter, and others will be highlighted along with references as to where to obtain additional information.

One of the areas in which major advancements have been made to concrete resilience is through the repair techniques available to increase both strength and ductility or to provide a return to the original properties after damage. The main repair techniques include wrapping the concrete elements in steel or fiber, crack repair with epoxy mortar or concrete, and larger repairs using epoxy concrete. The repair section of this chapter provides more detail along with references for additional information.

Addition of fibers to concrete can provide multiple benefits, including increased ductility, control of cracking, improved durability and toughness, improved freeze–thaw behavior, and reducing shrinkage effects. Fibers can be used in addition to traditional reinforcing or for some applications as a replacement.

Recent advances in 3D printing of concrete structures have great potential to improve the recovery time after an event. Structures for small homes have been printed in under 24 h, the raw materials are typically locally sourced and readily available, and labor requirements are, in general, lower than those for conventional construction. The flexibility of 3D printing in terms of the size and form of the resulting structure also allows for quick adaptation to the specific site and requirements of the structure. Deploying 3D printing technology could make replacing damaged structures or providing housing for emergency or longer-term workers cheaper and faster, helping to address the return to functionality part of the resilience equation.

Perhaps the most significant advances have been made in concrete mix design. Ultra-high-performance concrete, with strengths up to 29,000 psi (200 MPa), along with high-strength reinforcing, has allowed for thinner, lighter structures and the design of super tall buildings. Mix designs can also be customized to increase durability, freeze–thaw behavior, abrasion resistance, chemical resistance, and decreased environmental footprint. The improved performance is obtained through the use of supplemental cementitious materials (fly ash), additives, and controlling the aggregate and water/cement ratio.

1.3.3 Recommended Practices

One of the greatest advantages of using concrete is that standard design practices, properly applied for the conditions to which a structure will be subjected, produce a highly resilient project without taking extraordinary steps. However, several practices will help to maximize the objective resilience of a concrete design without any significant increase in cost:

- Detailing: Particular attention must be paid to detailing in the design. This is clearly true for loads such as seismic and blast, where large ductilities must be obtained under the design loads. However, the details for other loading conditions are equally important and go beyond requirements for reinforcing to include crack control, deformation compatibility, connection between elements and building/structure sections, and so on. Guidelines and standards are available for all these conditions and should be used in conjunction with good engineering practices to produce designs that will achieve the resilience goals of the project.

- Mix design: As discussed in the advances section, mix design sophistication has increased greatly, along with the ability to tune the mix to achieve the desired properties. By adjusting the mix, concrete can be produced that has any number of desired properties, including increased strength and/or ductility, reduced cracking, and greater durability. Tuning the concrete mix to meet the requirements of the project can increase the resilience in the desired ways.
- Anticipate loading changes: A major concern in society is the effect of climate change. It has been proposed that climate change may increase the frequency and severity of severe weather, leading to higher wind loads and greater flooding. In addition, rising sea levels will affect structures built near the coast. Anticipating these potential changes in demands on a project over its lifetime will allow for increased resilience for as long as it remains in service.
- Anticipate future uses: Not all structures retain the same function over their entire lives. Accounting for possible changes in occupancy will potentially increase the functionality of a project, allowing for increased resilience and better use of resources.
- Consider energy usage: Reducing the operational energy use of a building increases value for the owner and helps improve resilience. Buildings with high thermal mass can significantly decrease the energy requirements in certain climate zones. In addition to saving money during normal operation, the reduced energy requirements and retained heat/cold resulting from thermal mass also allow a building to remain functional for longer after an event if utilities are not available. The reduced energy usage also makes it easier to provide the services onsite through generators or renewable energy options.

REFERENCES

Abrams, M. S., and A. H. Gustaferro. 1968. *Fire endurance of concrete slabs as influenced by thickness, aggregate type and moisture*. Research Department Bulletin No. 223. Skokie, IL: Portland Cement Association.

ACI (American Concrete Institute). 2006. *Code requirements for environmental engineering concrete structures*. ACI 350. Farmington Hills, MI: ACI.

ACI. 2019a. *Code requirements for assessment, repair, and rehabilitation of existing concrete structures and commentary*. ACI 362. Farmington Hills, MI: ACI.

ACI. 2019b. *Building code requirements for structural concrete*. ACI 318-19. Farmington Hills, MI: ACI.

ASCE. 2011. *Blast resistant design of buildings*. ASCE 59. Reston, VA: ASCE.

ASCE. 2014. *Flood resistant design and construction*. ASCE 24. Reston, VA: ASCE.

ASCE. 2017a. *Minimum design loads and associated criteria for buildings and other structures*. ASCE 7-16. Reston, VA: ASCE.

ASCE. 2017b. *Seismic evaluation and retrofit of existing buildings*. ASCE 41-17. Reston, VA: ASCE.

Bilow, D. N., and M. E. Kamara. 2008. "Fire and concrete structures." In *Proc., Structures 2008: Crossing Borders*, edited by D. Anderson, C. Ventura, D. Harvey, and M. Hoit, 1–10. Reston, VA: ASCE.

FEMA. 2014. *Taking shelter from the storm: Building a safe room for your home or small business*. FEMA P-320. Washington, DC: FEMA.

ICC (International Code Council). 2014. *Standard for the design and construction of storm shelters*. ICC 500. Country Club Hills, IL: ICC.

ICC. 2018a. *International building code*. Country Club Hills, IL: ICC.

ICC. 2018b. *International existing building code*. Country Club Hills, IL: ICC.

ICRI (International Concrete Repair Institute). 2008. *Guideline for surface preparation for the repair of deteriorated concrete resulting from reinforcing steel corrosion*. ICRI 310.1-R. St. Paul, MN: ICRI.

Keremides, K., M. Oomi, R. Pellenq, and F. Ulm. 2018. "Potential of mean force approach for molecular dynamics-based resilience assessment of structures." *J. Eng. Mech.* 144 (8): 04018066.

NPA (National Parking Association). 2015. *Parking facility maintenance manual*, 5th ed. Washington DC: NPA

Texas Tech University. 2004. *Construction materials threshold testing*. Lubbock, TX: Wind Science and Engineering Research Center.

CHAPTER 2
STRUCTURAL STEEL AND RESILIENCE

John Cross, Charles Carter, Tabitha Stine, Luke Johnson

2.1 INTRODUCTION

Every group discussing resilience comes to the topic with its own perspective, develops its own definition, sets its own priorities, and drives the discussion down a different path. Perhaps the definition of resilience developed by the Rockefeller Foundation summarizes resilience best by stating "resilience means different things across a variety of disciplines, but all definitions are linked to the ability of a system, entity, community or person to withstand shocks while still maintaining its essential functions. Resilience also refers to an ability to recover from catastrophe, and a capability of enduring greater stress" (Rockefeller Foundation n.d.).

For the purposes of this chapter, resilience will be considered as the ability of an object or system to absorb and recover from an external shock, such as those caused by natural disasters (earthquakes, hurricanes, tornados, wildfires) or malicious acts (arson, terrorism) (Brand and Jax 2007).

Although the primary purpose of building codes is to protect the health and safety of occupants during an extreme event (Ching and Winkel 2016), the design goal of a resilient structure is the ability of the structure to withstand an extreme event with recoverable damage. By doing so, the building will either be able to maintain continuous function or be rapidly repaired and returned to service quickly.

The material selection for a building's structural framing system impacts the resilience of the structure by reducing the risk and the cost of that risk associated with the ability of the structure to absorb and recover from the stress of an extreme event. Of all the materials used for structural framing systems, structural steel has demonstrated the greatest level of resilience relative to extreme events. This is verified by significantly lower builder's risk and all risk premiums in the current insurance market for buildings framed with structural steel compared with concrete and wood. [Informational quote provided by Greyling Insurance Brokerage and Risk Consulting, Alpharetta, Georgia. Similar information relative to insurance rates for wood structures can be found in the *Claims Journal* and on the Steel Framing Industry Association website (SFIA n.d.).] The reasons for these lower rates and the greater resilience of buildings built with structural steel are structural steel's inherent durability, strength, elasticity, noncombustibility, and resistance to decomposition. The reduced rates are also benefited by the capability of structural steel framing systems to resist extreme loads, be rapidly repaired, and adapt to changing structural requirements.

2.2 IMPACT OF FRAMING MATERIAL SELECTION ON RESILIENCE

Resilience is a simple concept, yet it has complex implications for designers and builders. For some, resilience is viewed at the community level and refers to a community's ability to absorb and recover after a disaster. This can be measured by the ability to restore energy, transportation, clean water, and communication services to residents quickly after a disaster. Communities become resilient by having an infrastructure that includes buildings that can withstand intense storms or disastrous events. Facilities such as fire, police, health care, government entities, and designated shelters or residential units are of key concern for community resilience. At the same time, all the buildings in the community must be able to provide occupant safety during the event and some of those buildings must continue to provide critical services. In essence, resilience is the ability of a community, an infrastructure system, or a building to anticipate, prepare for, and adapt to changing conditions and withstand, respond to, and recover rapidly from disruptions. [H.R. 2241—Disaster Savings and Resilient Construction Act of 2013 (Pending Action).] Community resilience is built on the building blocks of its infrastructure, buildings, and essential societal services such as police, fire, health, and governance (Figure 2-1). Inherent in this view of resilience is the ability of critical infrastructure components to be resilient in their own right, maintaining or rapidly recovering functionality from disruptive events such as earthquakes, intense storms, coastal flooding, or terrorism (Chandra et al. 2011).

To enhance community resilience, key decision makers must begin by selecting materials that can efficiently and effectively be used in the design and construction of resilient framing systems for critical structures. When

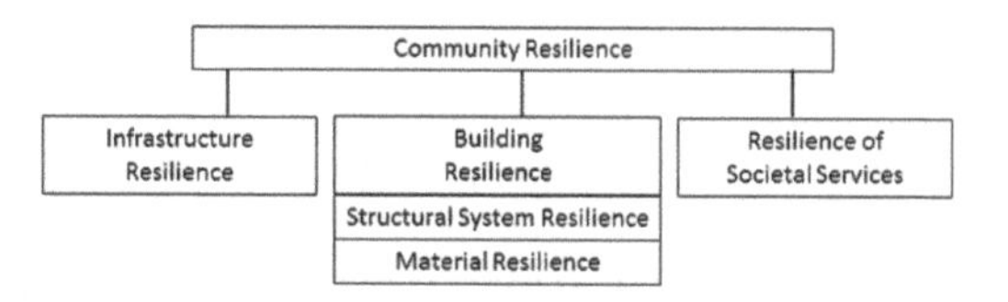

Figure 2-1. The building blocks of community resilience.

considering structural material and compared with other framing materials, structural steel clearly satisfies those requirements.

Any discussion of resilience becomes more complex when the discussion extends beyond the three building blocks of community sustainability to extreme events, often referred to as "stressors" that need to be accounted for. Natural events such as hurricanes, tornados, wildfires, earthquakes, flooding, and tsunamis are, in general, included, yet not all these stressors have the same likelihood of occurrence in every community; some may never occur in some communities. Events resulting from human activity including arson and terrorism also need to be considered. In some cases, technological events with no direct natural or human cause such as the faulting of an electric grid or the overloading of a communications gateway are included. Finally, the anticipation of future environmental events such as increased storm intensity, elevated water levels, and increased snow loads driven by global climate change may also need to be taken into account. Design for resilience is geographically determined and not all factors must be included when developing resiliency management plans and selecting building materials. Those "stressors" that have a measurable probability of impacting the local community must be considered. Clearly, any discussion of resilience is a multidimensional challenge combining discrete components, stressors, risk assessments, and future trends.

2.3 CHARACTERIZING RESILIENCE—THE 4Rs

Within the context of various definitions of resilience, the resilience of a community, building, or material is often characterized by four interconnected Rs: robustness, resourcefulness, recovery, and redundancy (AIA 2016).

At the community level, *robustness* refers to the ability of critical services to maintain operations during and after an extreme event. Buildings that house vital services such as health care, power management, transportation, and communications must be able to maintain operation for a community during and after a disruption. For a building to be resilient, it must also be robust and able to withstand or recover rapidly from an extreme event. An important component of a building's robustness is the integrity of the structural frame and, in turn, the strength of the material used in that frame.

Resourcefulness is the ability to prepare for and skillfully respond to a crisis or disruption. For a community, this means not only having contingency plans in place but also identifying and providing the resources needed to implement those plans. For a building, it means having "as-built" building plans available for rapid reference, structural engineers identified who are prepared to provide a rapid assessment of damage to the structural frame, and sources identified for materials that may be required to implement a repair. For example, structural steel is stocked at hundreds of steel service centers throughout the country for rapid delivery to a structural steel fabricator that can quickly fabricate the members required for the repair (AISC 2017b).

In the context of community resilience, *recovery* is the restoration of key operations as quickly and efficiently as possible after a disruption with the goal of a full return to normalcy within a short time frame. It is impossible and impractical to design a building and structural frame to handle every potential extreme event. There will be times when even the most resilient of designs is stressed to the point of failure. In these cases, resilience is determined by the level of loss of functionality and the time required to resume full functionality. The time required to accomplish a substantial recovery is in direct relationship to the robustness, redundancy, and ease of repair of the structural system, as well as the availability of resources to complete the repair.

Redundancy in the community context refers to the provision of backup resources to support key functional components of the resilient community. If a key component such as the provision of health services at a local hospital is taken offline, then a backup for that service should be identified to provide the service. For a building, redundancy best can be seen as the ability of the structural framing system and the material from which the frame is constructed to provide additional load-carrying capacity and the ability of the frame to transfer loads to alternate load paths.

Structural frames constructed using structural steel consistently receive high marks when measured using the 4Rs, thanks to the inherent resiliency of steel.

2.4 QUANTIFYING BUILDING RESILIENCE

A large number of design decisions will impact the resilience of a single building. Many of these design decisions will be in areas that have little, if any, connection to the structural framing system of the building. A resilient building may require redundant energy systems, secondary water supplies, particularly efficient heating, ventilation, and air conditioning

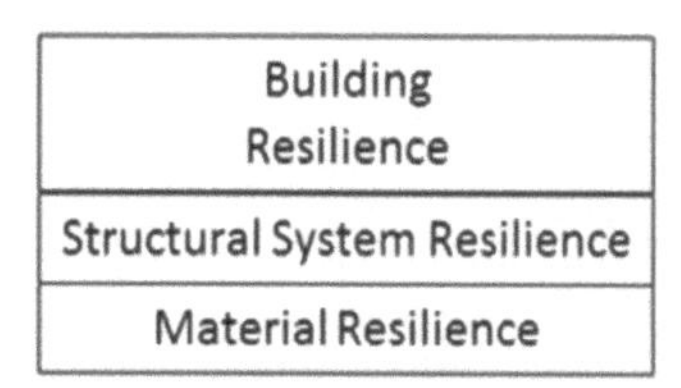

Figure 2-2. Structural components of building resilience.

systems, special glazing, and other specialized requirements; however, all these systems will be dependent on the resilience of the structural framing system. Bluntly put, if a building's structural system fails during an extreme event, the resilience of all other systems that are a part of that building will be of no value.

The key factor of a building's resilience is the resilient design of its structural framing system. The key consideration of the resilient design of a structural framing system is the resilient characteristics of the material selected for use in the framing system (Figure 2-2).

2.5 MATERIAL PROPERTIES

There is no single physical measure of resilience for a building material. When the resilience of a material is assessed, the primary attributes of the material must be evaluated. For a structural framing material such as structural steel, concrete, or wood, the attributes include durability, strength, elasticity, toughness, combustibility, and resistance to decomposition.

2.5.1 Durability

Durability is the ability of a material to withstand outside forces while sustaining minimal wear, fatigue, or damage. In Lewry and Crewdson (1994), several factors were identified that impact the durability of a product. These include weathering, stress, biological attack, incompatibility, and use. They also suggest that a contextual evaluation of these causes is the best way to determine the significant causes of degradation and that the best metric that could be used to measure durability is the service life of the product.

Service life estimates for framing system materials are available from a variety of sources but follow the same general pattern as indicated subsequently (Figure 2-3) (SCS Global 2013).

Steel had the highest marks in both nonresidential and utility pole construction when compared with concrete and wood.

Years	Non-Residential	Utility Poles
Steel	83	80
Concrete	81	60
Timber	69	50
Laminate	65	-
	costmodeling.com	IVL - SRI

Figure 2-3. Typical service life by material.

In addition, of the three materials, wood was ranked last in durability in a survey of 910 design and construction professionals conducted by FMI Management Consultants (2012). Although both concrete and steel were ranked highly, steel's durability was considered its leading benefit.

2.5.2 Strength

Steel is the strongest of the typical building materials. The design strength of most hot-rolled structural steel sections in use today is 50 ksi in both tension and compression, with special applications using sections with strengths as high as 70 ksi (MSC 2018). Compressive strength for concrete is typically between 3 and 5 ksi with some applications calling for high-strength concrete with compressive strengths as high as 15 ksi. Concrete tensile strength averages about 10% of concrete's compressive strength or is in the range of 0.5 ksi (Figure 2-4). The weakness of concrete in tension requires the addition of reinforcing steel in a building's beams and columns. The compressive strength of wood varies by the variety of wood, moisture content, and whether the load is applied parallel or perpendicular to the grain of the wood. Hardwoods have compressive strengths parallel to the grain in the range of 7 to 10 ksi (1 ksi perpendicular to the grain), whereas softwoods range from 5 to 8 ksi parallel to the grain (under 1 ksi perpendicular to the grain). The tensile strength of wood perpendicular to the grain averages about 1 ksi. While wood is relatively weak in tension perpendicular to the grain, it is strong in tension parallel to the grain exhibiting strengths in the range of 10 ksi (http://www.woodworkweb.com/woodwork-topics/wood/146-wood-strengths.html).

	Compressive Strength		Tensile Strength	
Structural Steel	50 ksi (as high as 70 ksi)		50 ksi (as high as 70 ksi)	
Concrete	5 ksi (High Strength 15 ksi)		0.5 ksi	
	Parallel to Grain	Perpendicular to Grain	Parallel to Grain	Perpendicular to Grain
Hardwoods	7 - 10 ksi	1 ksi	10 ksi	<1 ksi
Softwoods	5 -8 ksi	1 ksi	10 ksi	<1 ksi

Figure 2-4. Strength of materials.

The fact that the compressive and tensile strengths of structural steel are identical is a major factor in the ability of a structural steel framing system to resist and respond to extreme events. Unanticipated loads are often experienced by the structure in an extreme event, and in many cases, this is not just an increase in an anticipated load but rather the structural member unexpectedly transitions from being in compression to being in tension. Steel's equal ability to handle compressive and tension loads helps to mitigate any failure that may result from this condition.

A tragic example of an extreme event resulting in a building failure and significant loss of life was the collapse of the Murrah Federal Building in Oklahoma City, Oklahoma, resulting from a terrorist bomb blast. A FEMA study of the failure (FEMA 277) (FEMA 1996) concluded that several factors contributed to the cause of the progressive collapse, including the lack of continuous reinforcement in the concrete transfer girders and floor slabs, as well as the detailing of the concrete columns that did not provide the redundancy and ductility required for the additional demands on the columns (FEMA 1996). The National Institute of Science and Technology (NIST) later conducted a study that demonstrated that had the building been framed in structural steel, the ductility and tensile strength of an equivalently designed steel column would not have resulted in the failure of the critical column and progressive collapse of the building (FEMA 2010).

2.5.3 Strength Predictability

The importance of material strength as a factor of resilience is confined not to the strength alone but also to the predictability of that strength. Structural steel is a manufactured product complying with an ASTM standard specifying a minimum strength; therefore, when it arrives on the project site, it is at a predictable full strength. This is not the case with either concrete or wood. Concrete strengths are specified in the contract documents, a mix design is determined, and the material is placed in a wet state at the project site. The mix is typically designed to reach or exceed design strength 28 days after placement, which is verified by a testing service. During the 28-day period or following that period if the test specimen fails to reach the design strength, the structure under construction has a greater degree of vulnerability to the impact of extreme events (Carter 2014). Wood is even more problematic in that the strength of a single variety of wood can vary greatly based on moisture content, growth patterns, and the alignment of the member with the grain of the wood. This unpredictability is reflected in the large number of reduction factors applied to wood strengths during design. With steel, the capacity you want is the capacity you get.

	Modulus of Elasticity
Structural Steel	29 ksi
Concrete	≈ 1.5 ksi
Wood	≈ 3.5 ksi

Figure 2-5. Elasticity of materials.

2.5.4 Elasticity

Elasticity is the ability of a material to be deformed and return to its original shape and maintain its material properties. The greater the resistance to change, the greater the elasticity of the material and the faster it returns to its original shape or configuration when the deforming force is removed. In other words, elasticity is measured as the ratio of stress to strain. For a given stress (stretching force per unit area), strain (amount of deformation in the direction of the applied force divided by the initial length) is much smaller in steel than in wood or concrete, resulting in a higher modulus of elasticity and an enhanced capability for handling extreme loads without cracking or permanently deforming (Figure 2-5). Similarly, the ductility of a material such as structural steel allows for the redistribution of forces to provide an alternate load path or to accommodate displacements caused by extreme events (https://physics.info/elasticity/).

2.5.5 Toughness

Toughness is a measure of the ability of the material to resist permanent deformation, fracturing, and cracking. It can be best measured as the area under the stress–strain curve (Figure 2-6). To be tough, a material must be both strong and ductile. For example, brittle materials (like ceramics) that are strong but with limited ductility are not tough; conversely, very ductile materials (like copper) with low strengths are also not tough. To be tough, a material should withstand both high stresses and high strains. In general, strength indicates how much force the material can support, whereas

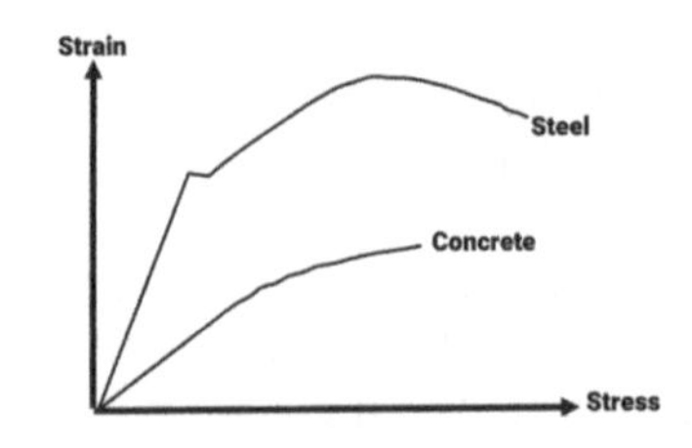

Figure 2-6. Typical stress–strain curves showing difference in area beneath curves.

toughness indicates how much energy a material can absorb before rupturing.

Steel is a much tougher material than concrete. Wood toughness varies greatly by species, water content, and the alignment of the grain, but even the toughest of wood does not achieve the same level of toughness as structural steel.

2.5.6 Combustibility

Materials that are "combustible" will burn, whereas materials that are "noncombustible" will not burn. Structural steel and concrete are classified by the International Building Code as noncombustible materials. Wood and timber are classified as combustible materials because they can burn (ICC 2015). Under extreme fire loads, concrete is subject to spalling, exposing steel reinforcement, and as a result, concrete's load-carrying capability will be reduced. To overcome the loss of load-carrying capability, an insulating covering may be placed around the structural steel to slow the loss of strength, allowing occupant departure and providing time for the fire to be extinguished. As the heat abates, the structural steel will return to its full strength, allowing the effects of the fire to be mitigated and the building returned to service. This is not the case with wood. Wood burns. Even if it is argued that wood simply chars, the cross-sectional area of the section is reduced, minimizing protection in the event of a second fire and reducing the cross-sectional area available to carry structural loads.

An unfortunate casualty of wood's combustibility was the Da Vinci apartment building in Los Angeles. In December 2014, the wood-framed apartment building, still under construction, burned to the ground (Figure 2-7). After the massive blaze consumed the building, the only feature that remained intact was the noncombustible steel-framed stairs (Los Angeles Magazine n.d.).

When a building built from wood burns, the building burns. When a building built from steel or concrete burns, the structure remains intact while the contents of the building burn.

2.5.7 Resistance to Decomposition

All materials are at risk of decomposition over time. Environmental factors such as humidity, moisture and air intrusion, mold, and mildew can cause a structure to deteriorate. When selecting a material, it is important to consider its resistance to decomposition in the climate zone where the construction will occur.

Steel and concrete, unlike wood, are inorganic and will not provide a source for mold, mildew, or structural deterioration (rot) to propagate. In

Figure 2-7. Postfire photo of the Da Vinci Complex showing only steel stairs surviving the fire.
Source: Courtesy of AISC, used with permission.

structures framed in either dimensional lumber or composite wood materials, rot can compromise the structural integrity of the structures, whereas mold and mildew compromise the health of the occupants.

One of structural steel's major advantages, when compared with other materials, is that steel will not absorb water in a flood situation or provide a moisture reservoir after the event. This is in contrast to concrete where all surfaces contain microcracks that can serve as paths for water to migrate to the reinforcing steel inside the concrete, resulting in corrosion of the steel and spalling of the concrete. Structural steel is not immune to the impacts of water inundation as corrosion on the surface of the steel may occur over time. Just as other materials may be able to mitigate the consequences of water inundation, corrosion can be prevented from occurring in locations where the structural steel may be exposed to flooding or other possible corrosive factors as paint or galvanized coatings can be applied that will provide protection. These coatings will provide protection for an extended period that will often exceed the anticipated service life of the structure. Corrosion detected on structural steel members during regular maintenance inspections is a surface condition that does not compromise the strength of the member. This corrosion can be addressed by cleaning the steel and applying a protective coating such as paint to the affected area (Troup and Cross 2003).

For a wood structure, decomposition can also be caused by pest infestation. Termite damage to buildings in the United States results in the loss of more than $5 billion annually (Orkin, https://www.termites.com/information/statistics/recent-statistics-about-termite-damage/). Structural steel and concrete are not subject to termite and pest damage.

In general, concrete decomposition such as spalling will require rehabilitation, wood decomposition such as rot will require replacement, whereas steel deterioration such as corrosion will require only maintenance.

2.6 INSURANCE CONSIDERATIONS AS A PARALLEL METRIC

Insurance policies are purchased by a building owner to cover damage, replacement costs, and loss of use of the building in the event of a disaster. Addressing resilience in the design of a building through a proper selection of the structural frame using appropriate structural framing materials is similar to purchasing insurance on the building (Wilson 2017). Spending money up front to address resilience can mean the difference between having a facility up and running shortly after a disaster or waiting months for reconstruction. The best measure of resiliency then becomes a measure of risk.

When determining how to quantify those risks, insurance rates based on today's market for Builder's Risk (insuring the building during construction) and All Risk (insuring the building after occupancy) are good indicators. Insurance companies regularly assess the loss records of buildings subject to both anticipated and extreme events and use those studies to set their rates. For a given set of risks, a lower rate means less likely damage and a lower cost of repair, providing an excellent proxy for comparing the resiliency of different structural framing materials.

The simple fact is that the insurance rates for structural steel–framed buildings are significantly lower than the rates for buildings framed in wood or concrete. Figure 2-8 illustrates typical insurance rates per $100 of value in today's market for Builder's Risk and All Risk insurance. These rates represent costs for the same building in the same location, with the only difference being the framing materials. According to Lauren Adami, CRIS, Greyling Insurance Brokerage and Risk Consulting, the rates for structural steel are consistently lower than the rates for wood or concrete, both during and after construction.

These rates are established based on costs for the same building in the same location and take into consideration the same external risk factors, including future impacts related to climate change. For example, in Florida, the risk could be based on calculations of a projected hurricane, storm surge, or flood damage, whereas in California, risk could be assessed based on calculations of projected earthquakes, wildfires, or mudslides. The rates change based on the project location and the particular risks associated with that locale and also depend on the project specialized features or aspects. Even though rates vary by location and project, the general trend is the same. Insurance rates for wood buildings are two to three times higher than those for an equivalent structural steel–framed building and

	Builder's Risk During Construction	All Risk After Occupancy
Wood	\$0.22 – \$0.27	\$0.20 – \$0.25
Concrete	\$0.14 – \$0.18	\$0.13 – \$0.16
Structural Steel	\$0.08 – \$0.12	\$0.08 – \$0.11

Figure 2-8. Typical insurance rates by framing material type.

the rates for a concrete building are 1.5 times higher than those for a steel-framed building. The difference is not the level of risk to the building from an extreme event, but rather the resilience of the building in response to that event. For a building valued at \$100 million, the savings in insurance costs over a 50-year period would be \$6.75 million for a structural steel–framed system when compared with those for a wood-framed system. The actuarial studies of the insurance companies confirm that structural steel–framed buildings are inherently more resilient than buildings framed in wood or concrete.

2.7 CONTRIBUTION OF STRUCTURAL SYSTEM DESIGN TO RESILIENCE

Structural framing systems can be designed to satisfy building code requirements using structural steel, concrete, and wood. The central purpose of building code provisions is to provide short-term human survivability and safety in the event of an extreme event. A competent structural engineer can accomplish this using structural steel, concrete, or wood. A more central question is not only meeting design goals because design goals can be accomplished using any of these materials, but the efficiency of using that material in the design, the cost of the system, the level of additional redundancy gained by the system, and the ease and speed of repair if the structural system is damaged in an extreme event.

2.8 BUILDING RESILIENCE

When selecting a structural framing system and material, there are many factors that must be considered.

2.8.1 Building Code Requirements

The resilience of a building's structural framing system is a function of both the material and the design of that system. Structural framing systems that meet the requirements of the building code can be built using steel, concrete, or wood. The purpose of building code provisions is to provide safety for occupants in the event of an extreme event. To ensure the best possible outcomes, all buildings are subject to building codes, such as IBC.

Building codes do not directly address the resilience of the building by referencing either a subjective evaluation of the 4Rs or an objective evaluation of the time and cost of repair and recovery. IBC in Section 1604 states that the design of building structures and parts of the building must consider strength, load and resistance factor, allowable stress, and empirical design or construction methods as permitted by the material.

IBC Section 1604.5 begins to address the resiliency of buildings by including enhanced design requirements for high-rise buildings in Risk Categories III and IV (ICC 2015). In these cases, structural integrity is evaluated independently. This means that deformations in the material are allowed, as long as failure does not occur. The goal is to allow for the redistribution of loads in the event of damage. This is possible to accomplish using structural steel, concrete, or wood. However, the question is not whether it is possible to accomplish this goal; the question is, what is the most efficient way to meet this goal by maximizing the design properties of the material specified in the design, getting the best return on investment from the system, achieving redundancy, and creating a system that is easy to repair after an extreme event.

When designing a structural framing system, there are two different philosophies on how to increase the structure's ability to handle extreme events. They are commonly referred to as a *bunker* philosophy and a *system* philosophy. The bunker philosophy attempts to handle an elevated level of potential loads by increasing the mass of the structure (i.e., build the bunker with thicker walls to handle larger forces applied to it). The increase in mass requires an increase in materials and an increase in cost. The system philosophy maximizes redundant load paths by using a material's natural ductility and reserve strength. During the design process, the system approach can integrate serviceability considerations. This allows for an emphasis on modularity and rapid repair.

A building designed following the bunker philosophy is not ideal because it is incredibly difficult to repair. By contrast, the system approach provides a technical solution that addresses the challenges of resilient design. A system solution would provide multiple options for lateral load resistance by utilizing a highly ductile environment that allows adequate member deformation, while still keeping access to critical services intact and operational. The design of a system with special connections and buckling restrained braces as structural fuses allow a structure to withstand an extreme event such as an earthquake or an event resulting from high winds or a blast. If damage occurs to the structural system, these components (the "fuses") can be efficiently removed and replaced, returning the structure to full functionality in a short period of time without major structure demolition or extensive retrofit.

It is not an efficient use of building materials if addressing the design requirements of high-risk buildings requires a bunker-style solution necessitating significantly increased material quantities. Increasing the mass of a structure, particularly a concrete structure, to address the challenges presented by extreme events is not an efficient solution. In contrast, structural steel supports a multitude of design approaches and innovative systems that address the challenge of resilient design from a technical rather than an increased mass perspective. Steel provides multiple options for lateral load resistance in a highly ductile environment that allows adequate member deformation, while still keeping access to critical services intact and operational.

An excellent example of a resilient structural steel framing system is the Tesla battery manufacturing facility, the Gigafactory, in Sparks, Nevada. The facility is designed with a fused rocking strongback frame that allows the lateral system to accommodate great variations in building configurations and equipment, while ensuring that the building will not collapse and be readily repairable in a 2,500-year earthquake. The system uses buckling restrained braces and Krawinkler Fuses for maximum energy dissipation, while the strongbacks and foundations remain elastic at full fuse yielding (Luth and Osteraas 2017).

In short, the bunker philosophy is a material-intensive solution and is not ideal for most applications, whereas the system philosophy provides a technical solution and is much more practical.

2.8.2 Design Redundancy

Unlike mix-dependent concrete or the variability of wood, structural steel provides additional redundancy and performs in a consistent and predictable manner as part of a structural system. Redundant load paths because of steel's natural ductility and reserve strength capacity provide additional structural capacity and resistance. In the design process, shapes are selected from a defined list; if load requirements fall in between two shapes, the larger section is selected, which provides additional strength beyond the basic design requirement. The design strength of the steel (F_y and F_u) is not the actual strength of the steel; the average actual strength of steel, which is greater than the design strength, can be estimated using the R_y and R_t multipliers found in the AISC Seismic Provisions (ANSI/AISC 341-40) (AISC 2016b). Although these values should not be used in routine design, they can be used to evaluate the resilience of the structure. Additional strength is also gained when beams are selected based on serviceability considerations of deflection criteria, floor vibration, or drift rather than the design strength.

2.8.3 Rapidity of Repair

To fully appreciate the required resiliency of a building is not only to assess the level of damage and the cost of repairs but also to consider the amount of time required to return the building to functionality. The time required to return to full operation is a function of the criticality of the services provided in the building, which should be taken into account in the initial design of the building. The return of a building to functionality may require the repair of the structural system, the replacement of structural components, and the temporary removal of portions of the structural frame to gain access to other building service components requiring repairs or replacement. Unlike concrete framing systems that would typically require demolition and replacement or wood systems that face the challenge of replacing numerous structural members after a flood or fire, structural steel can be strengthened in place through the use of doublers and stiffeners, structural members can be added, and beams can be penetrated to allow the addition of other services. All this can be done using materials that are readily available through a network of local steel service centers and fabricators.

2.8.4 Contributions of Structural Steel to Building Resilience

When determining resilience as it relates to the 4Rs, a defined resilience scale is key. Without an appropriate scale, it is difficult for decision makers to determine ways to efficiently and effectively compare alternate approaches. The evaluation of resilience is a developing field with proposed approaches falling under either a subjective or an objective methodology.

Subjective evaluation of resilience ranks alternate design approaches or plans against each of the 4Rs using a three-step ranking system: high, medium, and low. The alternate design approaches or plans are then assessed by the individuals involved based on their sense of the relative merit of each of the 4Rs and a final rating given to each.

The objective approach to quantifying resilience takes one of two forms. The first requires the individuals involved in assessing comparative proposals or plans to prioritize each of the 4Rs on a 1 to 5 scale (robustness and recovery might both be 5, whereas redundancy might be a 3, and resourcefulness a 2) and then rate the level of each of the 4Rs on a 1 to 10 scale. The priority level and rank are then multiplied for each R and added together for the overall result.

A more analytical approach is to consider risk levels, cost of recovery, and the length of time required for recovery for each type of extreme event. For example, if the probability of a 6.0 seismic event was estimated to be one occurrence every 50 years (0.02), the cost for repair to the building was estimated to be $10 million, and if the time of lost occupancy was 120 days at a cost of $50,000 per day, then the resilience rating would be 0.02 * ($10,000,000 + 120 * $50,000) = $320,000. The overall rating would be the sum of all possible events. Structural alternatives are then compared, with the lowest value indicating the greatest resilience.

While this approach may seem arbitrary and overly complex, it is exactly the same actuarial approach used by insurance companies to set insurance rates for buildings.

In addition to the aforementioned 4R-based method, the US Resiliency Council (USRC) has developed a rating system that addresses the following:

- Multiple dimensions including safety, damage, and downtime;
- Hazard levels consistent with current codes to assess safety, and possibly shorter return periods to evaluate damage and downtime;
- Rating symbols/indicators that communicate the level of performance in a simple effective way;
- Quality control and qualification system;
- Requirement that licensed engineers produce ratings for commercial buildings;
- Estimation of performance based on a defined, absolute scale rather than relative only to other buildings; and
- Cost structure reflecting the level of evaluation necessary to prepare the rating.

The rating system evaluates three categories: safety, repair cost, and recovery. Each category is based on a 1 to 5 scale with 5 being minimal impact in that category and 1 being maximum (Reis and Mayes 2015).

When the 4Rs, insurance rates, or the USRC results are looked at in the context of the use of structural steel as a structural framing material, the conclusion is clear—structural steel is an excellent choice as a structural framing material.

2.8.5 Robustness

Buildings framed with properly designed structural steel framing systems are robust and provide the necessary building skeleton for the maintenance of building operations in the event of an extreme event. The example previously cited of the Tesla facility is but one example. The structural steel–framed Torre Mayor tower in Mexico City is one of the tallest structures in Latin America at 55 stories and 774 ft. It was designed with viscous damper technology in a patented diamond configuration. One month after being occupied, it was subjected to a major seismic event, the 7.5 Tecoman, Colinas earthquake. Not only did the building survive undamaged, but the occupants did not even recognize that the earthquake had occurred (Rahimian et al. 2019). The robustness of the structural steel framing system, coupled with the innovative viscous dampers, provided the robustness required for the building to maintain all critical services with no interruption of normal functions.

2.8.6 Resourcefulness

There are times when structural steel buildings and bridges do experience damage from an extreme event. In those cases, it is critical that the resources be available to quickly address and repair the damage. Structural steel–framed structures address the issue of resources in two ways.

In the event of structural damage, it is critical to have rapid access to the as-built, dimensionally accurate structural plans of the structure that include not only the dimensions of the sections being used but also the characteristics of these sections. AISC-certified steel fabricators create and archive these material test reports on a piece-by-piece basis and project drawings and/or models for every project they complete. This allows repair activities to begin quickly and accurately.

Not only are plans available for structural steel–framed buildings, but the resources required to repair the structures are also readily available.

Material required for the repair of a structural steel framing system is stocked at service centers (steel warehouses) throughout the United States. Approximately 70% of the structural steel sections and plate that is purchased by fabricators for their regular use is purchased from a steel service center. These service centers stock between a 2- and 3-month inventory of structural sections (hot-rolled and hollow structural sections) and plate, totaling well over a million tons of structural steel to address project requirements (AISC 2018). This inventory is readily available for use in repair and reconstruction activities. Many of these service centers are a part of a larger national network allowing for quick acquisition of required material. Typically, material sourced from a service center can be delivered to a fabricator's facility within a 24 h period.

Behind these service centers stand the mill producers of hot-rolled structural sections, hollow structural sections, and plates that have demonstrated the capacity to meet any anticipated demand requirements of the current market. While mill deliveries typically take longer, as they are based on published rolling schedules of families' shapes, future rolling schedules will be modified to meet market demand experienced for certain sections at the service center level. In effect, the service center serves as a shock absorber in the structural steel supply chain, guaranteeing rapid delivery of sections and plates required for repair projects.

At the same time, there are more than 1,700 structural steel fabrication shops located in the United States. Among these 1,700 fabricators are 990 facilities certified for the fabrication of buildings and 355 facilities certified for the fabrication of bridges. The typical steel fabricator is a family-owned business employing anywhere from 10 to 100 people (AISC 2018). Projects may range from the fabrication of several tons of structural steel for a small retail store, to 25 ton for a rural overpass, to tens of thousands of tons for a large, high-rise structure or major bridge. These fabricators can address emergency repair projects quickly and handle increased workload requirements by working overtime, adding shifts, or subcontracting work to other fabricators. It should be noted that structural steel is a strong, light, and compact material that travels well. It is not unusual, for the distance from the fabricator to a job site to extend for hundreds, if not thousands, of miles. This immeasurable increases the pool of fabricators available to address emergency repair projects.

As previously noted, the fabrication process for a building is driven by plans developed and sealed by licensed structural engineers that specify all design aspects of the structural components. This design activity is required irrespective of whether the framing material is structural steel, concrete, or wood. For a steel-framed system, material acquisition, detailing, and fabrication will begin immediately on the receipt of the approved plans. Fabrication and erection costs for structural steel vary greatly based on the type of structure being constructed, the number of pieces, local labor conditions, and the complexity of the connections. An AISC member fabricator in the area of the project (a list of member fabricators is available on the AISC website—www.aisc.org) or the AISC Steel Solutions Center (866.ASK.AISC) is the best source for fabrication costs for a specific project.

Many fabricators, both AISC members and nonmembers, have taken the additional step of obtaining AISC Quality Certification. This program is similar to an ISO certification program but specialized for the intricacies of steel fabrication. Companies are audited on an annual basis; whereas the program does not certify a product, it does verify that the fabricator has the processes, equipment, manpower, commitment, and experience to perform the necessary work and meet a minimum level of industry-accepted quality standards. Currently, 990 fabrication facilities are building-certified, 353 are bridge-certified, and 139 hold bridge and highway component certification. Fabricator certification can be a recognized requirement for public projects, including almost all bridges. A list of AISC-certified fabricators is available at www.aisc.org/findcertification.

2.8.7 Recovery

The time required to recover from an extreme event is directly tied to the robustness of the structure and the resources available for the repair. The reality of the rapid recovery of structural steel–framed buildings and bridges is best illustrated by the example of the MacArthur Maze.

Figure 2-9. MacArthur Maze after failure.
Source: Courtesy of Lacy Atkinson/San Francisco Chronical/Polaris, used with permission.

The MacArthur Maze is a large freeway interchange at the east end of the San Francisco–Oakland Bay Bridge (Figure 2-9). On April 29, 2007, a tank truck carrying 8,600 gal. of gasoline overturned and caught fire beneath one of the ramps of the interchange. The petrochemical fire weakened the steel structure supporting the roadway, resulting in a collapse of the ramp connecting I-80 east to I-580. The original cost estimate for repair of the ramp was $10 million and a schedule that required the roadway to be out of service for several months, resulting in significant out-of-pocket costs to commuters and municipal agencies that provided free transportation on the local BART system. The State of California projected that the economic impact of the road closure was $6 million per day. Contrary to the initial cost and schedule estimates, the roadway was placed back in service on May 24, less than 30 days after the original accident at a cost below the original budget estimates (the actual winning bid was $876,075 with an incentive of $200,000 per day if the work was completed before June 27). This rapid recovery after an extreme event was accomplished because the material and labor resources required for completing the project were immediately available. Engineers were prepared to address the design issues on an accelerated schedule, a contractor with significant experience in rebuilding damaged expressways had an existing relationship with Caltrans, and the material (steel) and fabrication resources were readily available to the project team (Tradeline n.d.).

2.8.8 Redundancy

Unlike mix-dependent concrete or the variability of wood, structural steel provides additional redundancy and performs in a consistent and predictable manner as a structural system. Redundant load paths because of steel's natural ductility and reserve strength capacity provide additional structural capacity and resistance. The ability of structural steel–framed buildings to redistribute load has resulted in buildings with severe damage not failing or providing additional time for building occupants to exit the structure (Marchland and Alfawakhriri 2004).

Substantial research is currently being performed relative to the structural design of buildings to further improve the redundancy of structural steel framing systems, particularly in seismic regions (Simpson 2018). This work will further enhance structural steel's inherent redundancy and load-transferring abilities.

2.9 CONTRIBUTION OF STRUCTURAL STEEL TO COMMUNITY RESILIENCE

Obviously, the greatest contribution of structural steel framing systems to community resilience is the ability of a structural steel frame to withstand and recover from an extreme event, thereby keeping the building in service and benefiting the overall resilience of the community. However, beyond just maintaining building function, structural steel also brings other benefits to community resilience.

Extreme events that impact an entire community rather than just a single building generate significant amounts of waste. The majority of this waste is wood. Wood waste will be either burned or landfilled. Although some wood waste is reused or recycled in the normal construction cycle, it is most likely that the wood waste resulting from an extreme event will not be suitable for reuse. Burning or landfilling wood releases greenhouse gases into the atmosphere. Burning also generates particulate matter harmful to human health (EPA 2021). Landfilling requires sufficient landfill volume to be available to handle the increased flow of waste. While concrete may be crushed and down-cycled for use as a road base, a significant portion is also landfilled. Structural steel, on the contrary, is a fully recyclable material with an active market for its sale; it will not end up in landfills but will be returned to steel mills for recycling into new steel products. It will not be a burden on the community as the community seeks to rebuild (Figure 2-10).

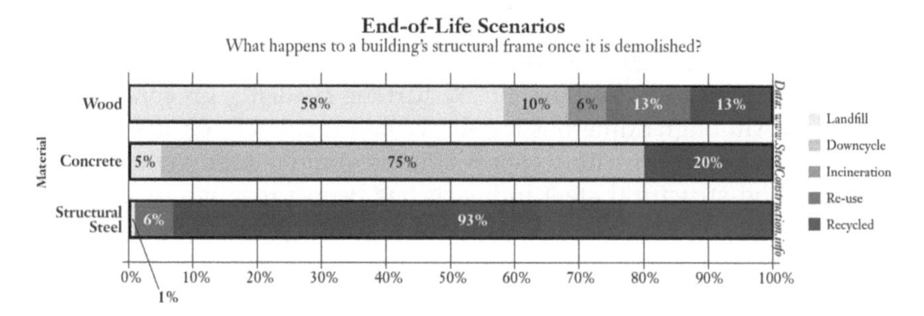

Figure 2-10. End-of-life destination by material.

When structures have to be renovated, remodeled, or rebuilt after a devastating event, utilizing a material that can be reused or recycled becomes beneficial from a cost, convenience, and sustainable standpoint. Materials, such as structural steel, that can be quickly retrofitted, replaced, and eventually recycled make a positive impact on the environment and community. A full 100% of deconstructed structural steel frames can be recovered and recycled for the production of new steel. Currently, domestically produced structural steel has an average recycled content of 93% and a recovery rate of 98%.

2.10 DELICATE BALANCE OF RESILIENCE AND SUSTAINABILITY

It would be shortsighted to end this discussion of structural steel and resilience without addressing the topic of sustainability. Although resilience and sustainability are separate topics and issues, a significant amount of overlap exists between the two. A building or a bridge that is not resilient is not sustainable. The demolition of the damaged structure will generate waste and the reconstruction of the building or bridge will require resources and energy, the production of which will generate environmental impacts. On the flip side, a building or bridge that is not sustainable cannot be considered resilient in that it will have a shortened life expectancy, higher maintenance costs, and harmful impacts on the community of which it is a part.

If increasing the resilience of a building means significantly increasing the mass of the framing system resulting in an increase in environmental impacts, then that solution is not a sustainable solution.

If increasing the sustainability of a building means selecting building products with lower environmental impacts but sacrificing robustness, resourcefulness, recovery, and redundancy, it is not a resilient solution.

Structural steel contributes not only to the resilience of buildings but to their sustainability as well.

Structural steel has long been considered the premier green construction material, and the structural steel industry continues to improve its leading environmentally friendly position by further reducing greenhouse gas emissions. Although numerous legislative and regulatory efforts in recent years have targeted emissions, energy efficiency, and related environmental concerns, the structural steel industry has been proactive in pursuing measures of its own that typically exceed regulatory requirements.

2.10.1 Recycled Content

The recycled content of structural steel can be as high as 100% for steel produced using the electric arc furnace (EAF) method of production. All domestic hot-rolled structural shapes are produced using the EAF method. A limited amount of virgin material (iron ore and alloys) may be added during the process to achieve the proper metallurgical balance required for a particular grade of steel resulting in an average recycled content of 93% for hot-rolled structural shapes. Hollow structural sections (HSS) are produced in a secondary process using hot-rolled coil formed into the tube shape. The hot-rolled coil can originate from either an EAF mill or a mill using a basic oxygen furnace (BOF). If the material is from an EAF mill, the recycled content will be in the 90% to 100% range. If it is from a BOF mill, the recycled content will be near 25%. Plate can also be produced in an EAF or a BOF mill, resulting in recycled content levels similar to those of HSS (AISC 2017a).

2.10.2 Recyclability

Irrespective of whether the structural steel originated from an EAF or a BOF mill, all structural steel is 100% recyclable. In fact, all steel products are recyclable. The steel used in an automobile can be recycled into the steel used in an appliance, which, in turn, can be recycled into a steel beam. It should be noted that this is true recycling without any loss of the material properties of the steel. Other materials are primarily down-cycled such as concrete into road base, not back into new concrete.

2.10.3 Recovery Rate

It is one thing if a material is recyclable, it is another thing altogether if the material is actually being recovered and recycled. By weight, 81% of all steel products reaching the end of their life are recovered for recycling. This includes 85% of automobiles, 82% of appliances, 70% of containers, 72% of reinforcing bar, and 98% of structural steel. The recycling rate for structural steel far exceeds the recycling rates of aluminum, paper, or wood. Recycling one ton of steel avoids the consumption of 2,500 lb of iron ore, 1,400 lb of coal, and 120 lb of limestone (AISC 2017a).

2.10.4 Reuse

Structural steel can not only be recycled but can also be reused without recycling. At the present time, only a small amount of recovered structural steel is refabricated and directly reused in new building projects. Appendix 5 of *The Specification for Structural Steel Buildings* (ANSI/AISC 360-16—available for free download at www.aisc.org/specifications) contains the testing requirements' evaluation of the properties for steel being recovered and reused (ANSI/AISC 360-16) (AISC 2016a).

A significant amount of structural steel is reclaimed from the waste stream of deconstructed buildings and industrial facilities for reuse in nonbuilding applications such as pipe racks, shoring, and scaffolding. Industrial steel structures are at times disassembled at one location for reinstallation and reuse at another location, an opportunity not available with other framing materials.

2.10.5 Waste Generation

All steel waste from the production, fabrication, or erection process is captured and recycled into new steel products. A recent survey of 900 steel fabricators indicated that not a single fabricator sends steel waste to a disposal facility. The rationale expressed by the fabricators was straightforward, "Why would we pay to send waste to a landfill when it is possible to sell the scrap to a dealer who will pick it up at our facility." Nonferrous waste generated in the fabrication process is minimal and limited to miscellaneous trash. All waste, including dust at the mill facilities with any ferrous content, is immediately recycled into the steel production process. Nonferrous waste is sold as by-products for other industries to use in their manufacturing processes. An example of the efficiency of the producing mills is highlighted at one facility where intact discarded automobiles are brought to the mill, shredded with the waste separated by material with the ferrous scrap flowing to the mill and nonferrous materials sold to waste processors, resulting in less than 1% of the original waste stream (the automobiles) transported to a landfill (Montalbo and Thinkstep 2017).

2.10.6 Water Consumption

Water consumption and discharge are minimal at mill facilities. While several thousand gallons of water are used to quench the molten steel sections (quenching not only cools the steel but also increases the strength of the steel), less than 70 gal. of water is actually consumed. The remaining water is recycled in a closed-loop recycling process and reused in the process.

2.10.7 Energy Consumption

The melting of steel scrap in an electric arc furnace is an energy-intensive process. The environmental impacts associated with the electricity required for this process are included in the life-cycle assessments and environmental product declarations for structural steel products. Unlike concrete in which a significant amount of greenhouse gas emissions occur as a function of the calcification process of the concrete, the greenhouse gas emissions associated with steel production are related to the external energy being used. This is significant as the environmental impacts associated with steel production are directly proportional to the emissions associated with electricity production. As renewable energy becomes a higher percentage of overall electricity production, the environmental impacts associated with steel production will decrease.

Interestingly, most structural steel mills utilize dispatchable energy contracts and attempt to schedule their melts to correspond with periods of low electricity demand using what could best be called "waste electricity." Waste electricity is the electricity being generated during nonpeak periods when coal-fired facilities cannot be easily cycled down to lower levels of generation because of the increased emissions that occur during the cycling process. The steel industry does not take any credit against the environmental impacts of structural steel for using this waste energy.

2.10.8 Offsite Fabrication

Sustainability is more than just inventorying environmental impacts. The triple bottom line of sustainability also includes both economic and social impacts. The fact that unlike wood or concrete in which only a limited amount of prefabrication takes place, all structural steel is fabricated in fabrication shops rather than at the project site. Offsite fabrication results in social benefits, including improved worker safety and a centralized work location, minimizing the requirement to travel to various project sites. With fabricated structural steel, the product goes to the project site rather than the workers. Although the erection of the structural steel frame requires a field crew, the size of a steel erection crew is significantly smaller than the number of workers required for stick-built wood or formed cast-in-place concrete structures.

2.11 ENVIRONMENTAL PRODUCT DECLARATIONS

An analytic presentation of the environmental impacts of structural steel can be found in the environmental product declarations (EPDs) published by AISC on behalf of the structural steel industry (EPDs for fabricated structural steel products are available at www.aisc.org/epd). A separate EPD is available for fabricated hot-rolled structural sections, fabricated hollow structural sections, and fabricated plates. These EPDs satisfy the requirements of various standards and rating systems, in that they document the environmental impact of these products as *delivered to the project site*. EPDs published by mill producers reflect only the impacts from the cradle-to-gate of the mill. The industry average EPDs, which are available at www.aisc.org/epd, document the environmental impacts from the cradle of the mill to the gate of the fabricator. The EPDs are based on life-cycle assessment data provided to an independent consultant for both mill and fabricator activities and have been peer-reviewed by a third-party reviewer.

The environmental impacts associated with a product are a function of the use of that product in a building system. The functionality of one ton of hot-rolled structural sections is not the same as the functionality of one ton of plate or one ton of hollow structural sections. The standards governing the development of life-cycle assessments and environmental product declarations require that language be included instructing the user not to use the data to make comparison across product categories (Figure 2-11).

These results are industry average values that include fabrication. Different mill producers may opt to determine producer-specific LCA values that will be published in producer-specific EPDs. These EPDs may or may not include fabrication impacts. Fabrication impacts are also calculated on an industry average basis and are determined by a detailed survey of over 300 AISC full-member fabrication firms.

However, the results for similar products can be instructive. For example, recent studies have shown that the global warming potential for hot-rolled structural sections produced in China is three times that of hot-rolled structural sections produced in the United States (AISC 2018).

It is important to realize that industry average values for fabrication give a clearer view of the environmental impacts associated with fabrication than could be obtained from fabricator-specific values. Each structural steel

All values are in kilograms per metric ton	Fabricated Hot Rolled Sections	Fabircated Hollow Structural Sections	Fabricatred Steel Plate
Global Warming Potential (GWP)	1.22E+03 CO_2 eq	1.99E+03 CO_2 eq	1.73E+03 CO_2 eq
Ozone Depletion (ODP)	1.63E-09 CFC-11 eq	1.62E-09 CFC-11 eq	1.62E-09 CFC-11 eq
Acidification (AP)	2.98E+00 SO_2 eq	4.34E+00 SO_2 eq	3.76E+00 SO_2 eq
Eutrophication (EP)	1.56E-01 N eq	2.34E-01 N eq	1.91E-01 N eq
Smog Formation Potential (SFP)	4.58E+01 O_3 eq	7.54E+01 O_3 eq	5.78E+01 O_3 eq

Figure 2-11. Domestic environmental impacts of 1 metric ton of fabricated structural steel (2020 data).

building project is a unique set of products (beams, columns, and so on) that require different levels of fabrication operations. If for the year under study a structural steel fabricator is involved in a series of steel-framed parking structures that utilize 60 ft beams that require minimal fabrication for connections, then the environmental impacts on a per ton basis will be low. If that same fabricator a year later is working on a specialized project with a large number of short sections requiring complex connections, then the environmental impacts on a per ton basis will increase dramatically. Fabricator-specific impact data are misleading, and no attempt should be made to select a fabricator for a project based on environmental impact data as these data are project-dependent (Cross 2018).

2.12 CONCLUSIONS

Material selection for a building's structural framing system impacts the resilience of the structure by reducing the risk associated with the ability of the structure to absorb and recover from the stress of an extreme event. The reduction of risk can be measured by an evaluation of the 4Rs (robustness, resourcefulness, recovery, and redundancy) and/or evaluating Builder's Risk and All Risk insurance premiums. Of all the materials used for structural framing systems, structural steel is the material of choice based on the application of those metrics, which has been demonstrated by consistent resilient performance in the face of extreme events. These lower rates and the greater resilience of buildings built with structural steel can be attributed to structural steel's inherent durability, strength, elasticity, noncombustibility, and resistance to decomposition. It is further aided by the capability of structural steel framing systems to resist extreme loads, be rapidly repaired, and adapt to changing structural requirements. At the same time, structural steel framing systems provide a proper balance between the sometimes competing desires of sustainability and resilience.

2.13 RECOMMENDATIONS

The selection of a structural framing material directly impacts the resilience of buildings and the communities of which they are a part. To effectively make that selection, the following recommendations are suggested for structural engineers and architects designing buildings and bridges:

- Evaluate the required level of resilience of the structure in the context of the community in which it is located.
- Identify the "stressors" prevalent in the area of the project.

- Determine key material criteria required to efficiently and effectively address the identified "stressors" including the sustainable characteristics of the material.
- Based on the selected material, develop a series of alternate framing system approaches to address the potential impacts of the "stressors" on the structural frame.
- If structural steel has been selected as the framing material for the project, engage a structural steel fabricator in the project area to gain insight into the fabrication cost of each alternative, the availability of materials both for initial construction and for potential repairs in the future, and the ability to access potential failure points within the structure in a rapid manner.
- Develop a detailed framing plan implementing the optimum approach to addressing the "stressors" present.
- Evaluation of the proposed structural system should be performed taking into account the actual average strength of the steel rather than the design strength and potential alternate load paths in the event of extreme loads being applied to the structure.
- Include with final project documentation all structural drawings, models, construction photos, and a report addressing the redundancy, availability of resources, and reparability of the structure. Ensure that as-built drawings are available, which will be very valuable in rapid fabrication in case of an extreme event.

REFERENCES

AIA (American Institute of Architects). 2016. "Building resiliency." In Chap. 3 in *Architectural graphic standards*. 12th ed. Hoboken, NJ: Wiley.

AISC (American Institute of Steel Construction). 2016a. *Specification for structural steel buildings*. ANSI/AISC 360-16. Chicago: AISC.

AISC. 2016b. *Seismic provisions for structural steel buildings*. ANSI/AISC 341-40. Chicago: AISC.

AISC. 2017a. *More than recycled content: The sustainable characteristics of structural steel*. Chicago: AISC.

AISC. 2017b. *Structural steel: An industry overview*. Chicago: AISC.

AISC. 2018. *Global warming potential of Chinese and domestic hot-rolled structural steel*. Chicago: AISC.

Brand, F. S., and K. Jax. 2007. "Focusing the meaning(s) of resilience: Resilience as a descriptive concept and a boundary object." *Ecol. Soc.* 12 (1): 23.

Carter, C. 2014. *Resilient—You can't even spell it without steel*. Chicago: AISC.

Chandra, A., J. Acosta, S. Stern, L. Uscher-Pines, and M. Williams. 2011. *Building community resilience to disasters: A way forward to enhance national health security*. Santa Monica, CA: RAND Corporation.

Ching, F. D., and S. Winkel. 2016. *Building codes illustrated: A guide to understanding the 2015 International Building Code*. Hoboken, NJ: Wiley.

Claims Journal. 2016. "Insurance costs 6 times greater for wood frame versus concrete." Accessed March 23, 2016. www.claimsjournal.com/news/international/2016/03/23/269612.htm.

Cross, J. 2018. *Unintended consequences*. Brooklyn, NY: Modern Steel Construction.

EPA (US Environmental Protection Agency). 2021. "Wood smoke and your health." Accessed November 1, 2021. https://www.epa.gov/burnwise/wood-smoke-and-your-health.

FEMA. 1996. *The Oklahoma City bombing*. FEMA 277. Washington, DC: FEMA.

FEMA. 2010. *Blast-resistant benefits of seismic design: Phase 2 study: Performance analysis of structural steel strengthening systems*. FEMA P-439B. Washington, DC: FEMA.

FMI Management Consultants. 2012. *Structural steel market assessment*. Raleigh, NC: FMI Management Consultants.

ICC (International Code Commission). 2015. Chap. 6, Sec. 1604 in *International building code*. Washington, DC: ICC.

Lewry, A. J., and L. F. E. Crewdson. 1994. "Approaches to testing the durability of materials used in the construction and maintenance of buildings." *Constr. Build. Mater*. 8 (4): 211–222.

Los Angeles Magazine. n.d. "Early takeaways from the Da Vinci complex fire." https://www.lamag.com/citythinkblog/early-takeaways-da-vinci-complex-fire/.

Luth, G., and J. Osteraas. 2017. "Delivering the GigaFactory in Tesla time using HD BIM." In *Proc., Structural Engineers Association of California, San Diego, California*. Sacramento, CA: Structural Engineers Association of California.

Marchland, K., and F. Alfawakhriri. 2004. *Facts for steel buildings: Blast and progressive collapse*. Chicago: AISC.

Montalbo, T., and Thinkstep. 2017. *Hot-rolled structural sections: Life cycle inventory method report, issued to American Institute of Steel Construction in July 2017. In appendix to China, global warming and hot-rolled structural steel sections*. Accessed November 1, 2021. https://www.aisc.org/globalassets/aisc/publications/white-papers/global-warming-potential-of-chinese-and-domestic-hot-rolled-structural-steel.pdf

MSC (Modern Steel Construction). 2018. *IDEAS2 awards*. Chicago: MSC.

Rahimian, A., R. E. Valles Mattox, E. Anzola, I. Shleykov, S. Nikolaou, and G. Diaz-Fanas. 2019. "Structural design challenges for tall buildings in Mexico City." *Structure*, February 2019.

Reis, E., and R. Mayes. 2015. *USRC building rating system for earthquake hazards*. Washington, DC: USRC Resources, US Resiliency Council.

Rockefeller Foundation. n.d. "Our work." Accessed April 2017. https://www.rockefellerfoundation.org/our-work/topics/resilience.

SCS Global. 2013. *Environmental life cycle assessment of southern yellow pine wood and North American galvanized steel utility distribution poles*. Emeryville, CA: SCS Global.

Simpson, B. 2018. *Sturdy spine*. Chicago: Modern Steel Construction, American Institute of Steel Construction.

SFIA (Steel Framing Industry Association). n.d. Accessed November 1, 2021. https://sfia.memberclicks.net/assets/FactSheets/insurance%20savings%20with%20cfs.pdf

Tradeline. n.d. "Unprecedented teamwork repairs collapsed freeway in record time." https://www.tradelineinc.com/reports/2007-10/unprecedented-teamwork-repairs-collapsed-freeway-record-time

Troup, E., and J. Cross. 2003. *Open deck parking structures*. Chicago: AISC.

Wilson, A. 2017. "20 Ways to advance sustainability in the next four years." *BuildingGreen*, 26(1): January.

CHAPTER 3
MONITORING FOR RESILIENCE IN HIGHWAY BRIDGES

Raimondo Betti

3.1 INTRODUCTION

3.1.1 Resilience Definitions

At a time when major environmental catastrophes such as hurricanes, floods, earthquakes, tsunami, and so on seem to repeat on a yearly basis and to affect larger sectors of our society, it has become of paramount importance to address the issue of the resilience of our transportation system. Defining resilience is already a difficult task on its own: since one of the earliest definitions of resilience given by Manyena (2006) as the "intrinsic capacity of a system, community or society predisposed to a shock or stress to adapt and survive by changing its non-essential attributes and rebuilding itself," there have been many attempts at providing a definition of resilience that is general enough to include all possible situations, from a single element (e.g., a bridge or building) to a network of elements or system (e.g., state-owned bridges), to multiple systems (e.g., transportation infrastructure, including roads, railroads, airports), to entire communities. Currently, there are two widely used definitions of resilience; see the National Infrastructure Advisory Council (NIAC 2009) and the National Security Council (NSC 2011). In the 2009 report titled "Critical Infrastructure Resilience," the National Infrastructure Advisory Council defines infrastructure resilience as "the ability to reduce the magnitude and/or duration of disruptive events. The effectiveness of a resilient infrastructure or enterprise depends on its ability to anticipate, absorb, adapt to, and/or rapidly recover from a potentially disruptive event." Such a definition of resilience is linked to three critical components:

1. *Robustness*: ability to maintain critical operation and functions during crises,
2. *Resourcefulness*: ability to skillfully prepare for, respond to, and manage a crisis or disruption as it unfolds, and
3. *Recovery*: ability to return to and/or reconstitute normal operations after a disruption.

The definition of resilience that the National Security Council provides in its 2015 National Preparedness Report is along the same lines as the one of the NIAC, defining infrastructure resilience as "the ability to adapt to changing conditions and withstand and rapidly recover from disruption due to emergencies." With this goal in mind, five different mission areas were identified:

1. *Prevention*: capabilities necessary to avoid, prevent, or stop an immediate threat,
2. *Protection*: capabilities necessary to secure the homeland against acts of terrorism and man-made or natural disasters,
3. *Mitigation*: capabilities necessary to reduce loss of life and property by lessening the impact of disasters,
4. *Response*: capabilities necessary to save lives, protect property and the environment, and meet basic human needs after an incident has occurred, and
5. *Recovery*: capabilities necessary to assist communities affected by an incident to recover effectively.

These two definitions differ in their main categorizations of the components that fall under the umbrella of resilience: while the former includes only three main resilience components, the latter includes five main resilience components. However, the subcomponents of both definitions are basically the same.

Recently, Ettouney and Alampalli (2011) provided a more comprehensive definition of resilience by introducing a fourth critical component to those used in the NIAC (2009) definition. They augmented the three critical areas of robustness, resourcefulness, and recovery with

4. *Redundancy*: ability to have backup resources to support the originals in case of failure.

This is an extremely important component, not highlighted in the other two definitions, for quantifying resilience in civil infrastructure systems. In fact, such systems, for example, either a single bridge or a network of bridges, are designed with a high level of redundancy, but because of a variety of factors linked to environmental and aging conditions, such redundancy level tends to decrease over time. Hence, it becomes extremely important to account for the current level of redundancy when evaluating the resilience of a single structure or of an infrastructure system.

Such a definition, including robustness, resourcefulness, recovery, and redundancy, is also known as the "4Rs" of resilience, after Ettouney and Alampalli (2011), and it is the one adopted in this chapter. It is obvious from this definition that talking about the resilience of a structure or of an infrastructure system (or even of a network of infrastructure systems) requires a multidisciplinary approach where engineering solutions are evaluated through economic and societal lenses.

3.1.2 Monitoring of Highway Bridge Resilience Components (4Rs)

In this chapter, we will focus our attention on the importance of monitoring in the assessment of the resilience of highway bridges, a component that is fundamental within the robustness aspect of resilience, because it is through monitoring the structural performance in operational conditions and/or during single extreme events that it is possible to assess the current structural conditions necessary to estimate the system's robustness and redundancy, the allocation of human, technical, and economic resources, and the potential recovery time.

Roadway bridges represent the backbone of the US transportation infrastructure. With more than 610,000 road bridges that carry more than 90% of the nation's supply chain, the roadway infrastructure is vital for the well-being and prosperity of our society. However, almost 40% of these bridges are approaching the 50 year mark of service life, whereas about 10% of the entire road bridge inventory is considered to be "structurally deficient" (ASCE 2017). According to the 2017 ASCE Infrastructure report card, the estimates of the backlog of the bridge rehabilitation needs approach $123 billion. With such a large number of bridges approaching their service life, such bridges become vulnerable not only to extreme events, as originally intended, but also to regular daily operations. For example, increasingly heavy truck loadings have been found to directly affect the service life of highway bridge superstructures. Damage typically occurs in the bridge deck and the main superstructure elements, including floor beams and girders, diaphragms, joints, and bearings. With the rapid growth of highway transportation, the increasing frequency of passing heavy trucks contributes to fatigue damage.

Another factor that strongly affects the resilience of highway bridges is the aging of such bridges, with consequent deterioration of the mechanical properties of the materials. Over the course of decades of service life, corrosion of web and flanges in steel bridges and rebars in concrete bridges, often undetected when they occur in inaccessible areas of the structure, has dramatically reduced the capacity of such structures. As a result, bridge maintenance becomes increasingly difficult and costly, because maintenance, rehabilitation, and/or replacement become more frequent.

The chapter will first survey current monitoring techniques for highway bridges. We then explore how some of these monitoring processes can be applied to assess two of the resilience components: robustness and redundancy, in real time (before, during, or after an event). We will then show how those same monitoring processes can be used to enhance recovery efforts. First, we will discuss recovery enhancement for situations where recovery resources (human, equipment, etc.) are located only in a single location. Then, we will show how monitoring techniques can be used to optimize resource utilizations in the case of more than one location for recovery resources.

The chapter concludes by offering some recommended best practices for the stakeholders of highway bridges.

3.2 STATE-OF-THE-ART IN MONITORING HIGHWAY BRIDGES

Roadway bridge infrastructure is among the most critical infrastructure systems in a modern society and, in particular, in the United States. Any disruption of service in highway systems in metropolitan areas because of repair or rehabilitation work can have severe repercussions on the population not only of the city itself but of the entire region, and, therefore, roadway bridge infrastructure must be carefully planned ahead of time.

In assessing the resilience of such a complex system, the importance of correctly assessing the robustness of each single bridge in its entirety as well as that of its components is evident. In addition, because the roadway bridge infrastructure system is an intricate network of regional, state, and federal networks, the resilience of the network itself at these different levels must be considered, bringing to the table different local, state, and federal stakeholders, with the understanding that it is the weakest link (either a component in a single bridge structure or a bridge in a network of bridges) that could alter the resilience assessment of a single bridge and/or of a network of bridges.

It is then clear that the difficult task of determining the resilience of a bridge (or of a network of bridges) is strongly dependent on having a proper assessment of the current structural conditions, not of those at the initial construction stage, and the first step to achieve this is to have a reliable and properly maintained monitoring system. Today, thanks to the advancements in sensing and computing technology, it has become relatively inexpensive to install a monitoring system on a critical structure. Bridges of a certain importance, for example, the Stonecutters bridge in Hong Kong or the Jindo Bridge in Korea, have hundreds of sensors permanently installed on different parts of the bridge and these sensors have been operational for a few years. This trend is slowly diffusing among state and local bridge owners/engineers who, for example, are now replacing (or augmenting) visual inspections with drone-assisted inspections to allow visual access to remote parts of the bridge structure. However, there are still barriers, mainly cultural, that prevent a full spreading of monitoring initiatives to the broad transportation infrastructure system.

Today, there are a variety of sensors/technologies that have been applied in monitoring bridge structures. In Table 3-1, some of these sensors/technologies are listed, also indicating the type of application with which they can be used. Such technologies can be used either as stand-alone or in conjunction with other technologies. For example, acceleration time histories recorded at specific locations on bridges offer insight into global bridge behavior, whereas strain gauges and/or acoustic emission sensors provide information on cracking development in specific structural components. However, while these monitoring systems can generate a large amount of data, there are still difficulties in translating this voluminous information into data that can be used by bridge owners in their resource allocation and decision-making process.

Few sporadic examples of systematic monitoring programs on highway bridges exist. An example is represented by the California Strong Motion Instrumentation Program of the California Department of Conservation, where a network of about 60 highway bridges has been permanently instrumented with accelerometers. Being in active seismic areas, bridges in California are frequently subjected to ground motion excitation of different magnitudes, providing a wealth of data that can be analyzed and used for structural assessment implementations. The data collected by these sensors were used by Mosquera et al. (2012) to quickly assess the structural integrity of a network of highway bridges after a major seismic event. Having the possibility of analyzing records of the ground shaking and of the bridge response at different locations and for events of different magnitude allowed the researchers to develop strategies to determine and quantify structural damage almost immediately after a new seismic event and to use this information for a more effective emergency and recovery plan.

Over the years, many highway bridges have been instrumented with sensors on a temporary basis. These sensors, mainly accelerometers but also strain gauges, tiltmeters, thermometers, and so on, were installed on a given bridge and removed after a certain period of time. A classic example is the old I-35 bridge in Minneapolis that collapsed in 2007: before its collapse, this bridge had some strain gauges installed at some critical locations that were in use for some time. Today, the new I-35 bridge has a permanent network of sophisticated sensors that record the structural response and environmental variables such as temperature at different

Table 3-1. List of Sensors/Technologies that can be used in Monitoring Highway Bridges.

Type of sensor/ technology	Type of measurement	Possible use
Accelerometers	Acceleration (in one or three orthogonal directions) at the sensor location	To update finite-element models or to develop data-based statistical models
Strain gages	Elongation/contraction at gauges location	To update finite-element models, to assess crack propagation in structural components
Tiltmeters	Small changes on inclination from the horizontal/vertical levels	To measure foundation settlements and/or rocking of the piers
Thermistors	Temperature	To account for thermal effects on structural components
Linear variable differential transformers	Linear position	To monitor relative displacements between a fixed point and targets on structural components
Laser scanning (LiDAR)	Noncontact measurement of the distance to a target	To monitor bridge settlements as well as cracking of structural components and decks
Global positioning system	Noncontact measurement of the position of a target on the structure	To monitor displacements of target points on the structure as well as traffic conditions
Digital cameras	Videos/photographs of the motion of a target on the structure	To monitor displacements of target points on the structure as well as traffic conditions
Acoustic emission	Energy released by crack openings	To monitor crack propagation in steel structural components
Resistivity meter/half-cell potential	Resistivity/electrical potential field in concrete	To monitor corrosion of rebars in concrete decks

locations (Linderman 2019). Another example of a sophisticated sensor network installed on a roadway bridge is the one presented in the work by Fraser et al. (2010): a 90 m long highway bridge in California was instrumented with an accelerometer sensor array and an integrated camera monitoring framework. Twenty accelerometers were mounted on the deck, along the length of the bridge, while a high-resolution camera was installed up a light post at the end of the bridge. With the data collected by this system, they were able to correlate the bridge deflection with the traffic loading.

Nowadays, new developments in digital photography are making it possible to obtain measurements of the structural response at various locations just by processing frames from videos recording by digital cameras. With just one digital camera properly placed in the proximity of the bridge, it is possible to record the time histories of displacements at as many target locations along the bridge from just one video. This will make the data collection phase very economical and easy to perform. As an example of vision-based monitoring of highway bridges, Hou et al. (2002) have used video recordings along a 20 mi highway corridor to integrate the measurement of vehicular loads with data from structural health monitoring (SHM) systems installed on two bridges so as to correlate the peak bridge response to trucks' weight.

Modern iconic bridges, for example, cable suspension and cable-stayed bridges, are being built today with extensive instrumentation systems already installed, systems that provide a large amount of structural and environmental data (Lus et al. 2002, Hong et al. 2011). In the past, isolated monitoring campaigns were conducted on such bridges: usually, accelerometers were placed on bridges for a limited period of time and data were collected and analyzed. It is noteworthy that, thanks to these iconic bridges, structural monitoring has undergone tremendous advances, both in terms of hardware (sensors) development and in terms of computational algorithms. However, a discussion on these bridges is outside the scope of this chapter and, therefore, will not be considered any further.

3.3 ROBUSTNESS/REDUNDANCY—TRADITIONAL STRUCTURAL HEALTH MONITORING

As previously mentioned, determining the resilience of a structural system or of a network of systems requires the assessment of the current conditions of the system. To this end, the first step in this direction is to monitor the system's performance during service operations: once the monitoring system has been installed on a bridge, measurements of the structural response (displacements as well as accelerations, elongations, and so on) at different locations are recorded during regular operational conditions and during extreme events such as earthquakes or windstorms. These data, usually time histories of the structural acceleration and/or displacement, are then processed by the engineer to provide a more or less accurate description of the bridge in its current condition. It is a sort of "system identification" in which, chosen a certain model as representative of the structure in question, its parameters are identified based on the information provided by the recorded data. Once this information becomes available, the engineer can then assess whether the bridge in question has suffered any substantial damage, or it is still in satisfactory working condition.

All the identification methodologies can be classified into two major categories, depending on the way they operate on the measurement data: (1) Those methods that perform their analysis directly in the "time" domain, for which natural frequencies, damping ratios, and mode shapes are by-products of the identification process; (2) those methods that translate the analysis into the "frequency" domain, where the natural frequencies, damping ratios, and mode shapes are the primary parameters of the identification process. Another classification of these methodologies is based on whether the methodology identifies a model that is characterized by physical parameters, such as stiffness and mass, or the model is described by a mathematical expression whose unknown coefficients have no physical meaning: obviously, from the point of view of structural condition assessment, the "physics-based" models are preferable because, in bridge engineering practice, it is usually accepted that structural damage is linked to a loss of stiffness (e.g., cracking, spalling, scour, and so on), and so, such models directly provide useful indication on the occurrence and location of damage.

One of the most common approaches for the identification of a reliable model of a highway bridge is called "model updating" in which an initial model of the bridge, usually a finite-element model, is created and updated using the results of the recorded data analysis. Although there are various forms of model updating, the updating process is always conducted following this general scheme (Figure 3-1):

1. First, the data recorded in the field are analyzed; for example, from the time histories of the structural response, the power spectral density functions can be obtained and characteristic structural parameters such as natural frequencies, damping ratios, and mode shapes are extracted.
2. At the same time, an initial FE model of the structure is created, using all available information from blueprints, photos, and construction records, and the same structural parameters (e.g., natural frequencies, damping ratios, and mode shapes) are obtained from it.

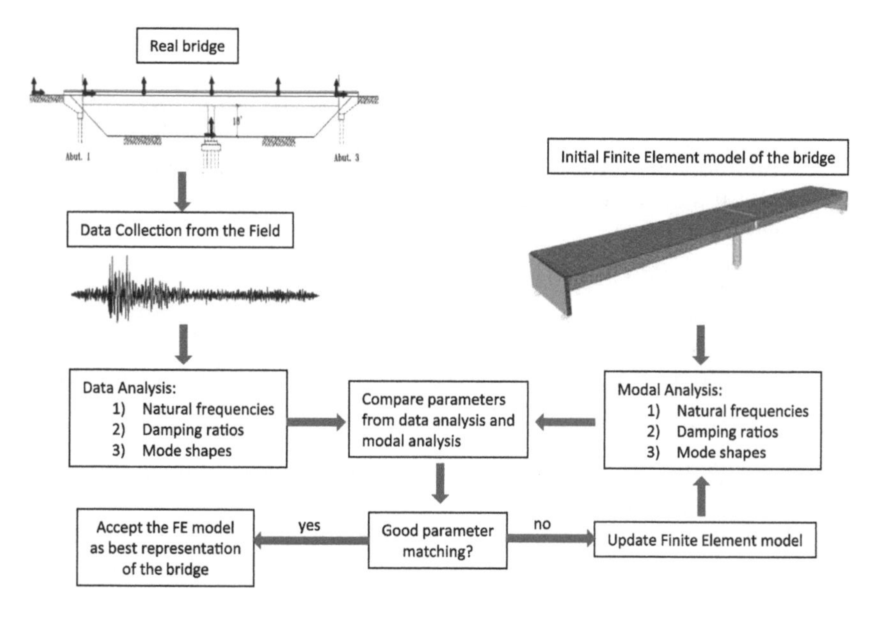

Figure 3-1. Finite-element model updating: Overall strategy.

3. Because, almost certainly, there will be a mismatch between the parameters obtained from the data analysis and those derived from the numerical model, some material and/or geometric characteristics (e.g., Young's moduli of the different materials, moments of inertia and areas of beams and columns, etc.) of the FE model will be adjusted so as to minimize the mismatch between the estimated parameters and the initial FE model parameters. At this point, an updated FE model of the structure will be obtained.

This is an iterative operation that requires the definition of the functional operator that needs to be minimized: this functional usually contains, in different forms and combinations, the errors between recorded and computed natural frequencies and mode shapes. Such operation is basically an optimization process that will stop when the functional reaches a minimum point, indicating that the updated FE model has now dynamic characteristics (natural frequencies and mode shapes) that are quite close to those obtained from the data analysis. This iterative minimization process follows rules borrowed either from the mathematical world (e.g., the steepest gradient) or from the "natural" world (e.g., genetic algorithm, artificial bee colony, etc.). One of the main drawbacks of this type of approach is that the iterative process might stop at a relative (not absolute) minimum point: in this way, even though no further minimization of the functional can be achieved, the FE model is not necessarily representative of the real structural system. Therefore, special attention has to be paid to avoid this kind of intermediate solution.

After converging to a reasonable matching of the parameters, the final FE model is now deemed ready to represent the structure in its current condition and can be used for testing the structural performance against several limit states. For example, for a bridge built in an active seismic area, the FE model could be tested for limit states such as the following:

a. Its own weight only, DL;
b. Nominal live load condition, $DL + \alpha LL$. The factor α can be chosen by bridge officials, as deemed reasonable; and
c. Emergency and/or recovery equipment, ERL condition, $DL + ERL$.

Then, a decision can be made about the bridge being

a. Not safe for any traffic (or even failed) if it does not meet the under the DL limit state;
b. Not safe for vehicles, if it does not meet the $DL + \alpha LL$ limit state; or
c. Not safe for an emergency/recovery equipment if it does not meet the $DL + ERL$ limit state.

Among the time-domain approaches used in identifying the structural properties, some algorithms that have been successfully used in the analysis of bridges (including highway bridges) are based on the identification of "black-box" models. The names of such models derive from the fact that the parameters describing such models have no physical meaning, and so, they cannot be directly used to determine whether and where damage has occurred. However, such models provide excellent input–output mapping, and so, they are capable of reproducing the structural response to an external excitation very accurately. Although quite different from one another in terms of the intermediate steps, all these identification methods follow the same basic idea, that is, to project the recorded time histories of the structural response on a vector space whose unit vectors have some given properties. For example, some methods provide a least-square solution (an orthogonal projection), whereas others use the geometric tools of oblique projections on the recorded data sets to obtain a set of matrices that, when multiplied by the input time histories, provide an excellent estimation of the structural response. As an application of such algorithms to bridges, Lus et al. (2002) made a first attempt to use such computational techniques to identify a linear, first-order model ("black-box" model) of the Vincent Thomas bridge in Los Angeles, California, using the measurements of the ground shaking and of the structural response recorded at various locations on the superstructure recorded during the Whittier earthquake. The identified matrices have no apparent physical meaning, but they provide an accurate estimation of the natural frequencies, of the damping ratios, and of the mode shapes at the recorded stations. In addition, such matrices could be used to estimate the maximum structural response at the monitored locations for future ground shaking once the time history of the ground motion for a future earthquake can be estimated.

Using the same identification methodology applied by Lus et al. (2002), Mosquera et al. (2012) proposed an interesting framework to rapidly assess the robustness of a network of highway bridges. They used low-level vibration measurements recorded on a multitude of highway bridges to determine high-fidelity first-order models of such structures using an algorithm called observer/Kalman filter identification (OKID). This is a quite efficient algorithm that requires no initial estimation of a model (contrary to the FE model updating procedure) and relies only on the recorded data of the input excitation (ground shaking) and of the output (structural accelerations and/or displacements and/or strain gauges) to determine a linear model of the structural system under consideration. Using only the time histories of the measured input excitation and structural acceleration at various locations, this algorithm provides a high-fidelity, linear black-box model that maps the input excitation into the output space. Having previously identified one of these models for each of the bridges in the network using the low-magnitude earthquake data available, when the record of the next seismic ground shaking becomes available, it is fed through the various models (already available to the engineer) and, in a very short time, the structural response at the different sensor locations is predicted for each bridge with the underlying assumption that the structure remains in the linear elastic range. If the predicted time histories of the structural response match the time histories recorded during the latest shaking, then the bridge can be considered undamaged (it remains within the linear range of response), and so, it can be used for immediate operations. Figure 3-2 shows the application of such a framework applied

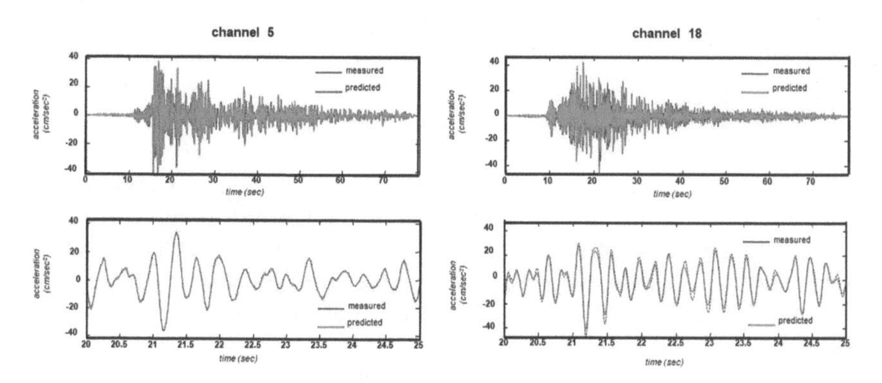

Figure 3-2. Comparison of the measured and predicted acceleration time histories for the El Centro-I8/Meloland overpass.

to the El Centro-I8/Meloland overpass, using the recorded structural accelerations and ground motion from the Cerro Prieto February 6, 2008, earthquake. A comparison of the time histories of the measured accelerations at Channels 5 and 18 and those predicted using a black-box model identified using the data from the Calexico, December 27, 2008, earthquake shows a perfect matching between the two datasets. It could be concluded that the model identified using the dataset from the Calexico earthquake accurately predicts the bridge response during the Cerro Prieto earthquake, and so, it can be assumed that no major structural damage has occurred.

On the contrary, if there is a mismatch between the recorded and the predicted structural responses, such a discrepancy could be indicative of the occurrence of structural damage, and in such a case, more in-depth analyses are required. At this point, it would be necessary to evaluate an updated FE model of the specific bridge that shows such a mismatch so as to identify the level of damage and its specific location. In the work by Mosquera et al. (2012), the FE model updating was accomplished by using a genetic algorithm approach and selecting the elastic modulus and unit weight of the concrete and the damping ratios as updating parameters.

In dealing with highway bridges, it is of great interest also to monitor the performance of such bridges in operational conditions: these are conditions that correspond to the regular everyday life use of the bridge and are not related to a particular event (e.g., earthquake, hurricane). In dealing with operational conditions, it is practically impossible to monitor the external excitation, and so, identification methods have to rely only on the measurements of the structural response. This limitation is quite restrictive because of the lack of information of the input excitation and prevents the engineer from identifying a full model of the structure. However, a "partial" black-box model can still be identified from which it is possible to extract information about the natural frequencies, damping ratios, and mode shapes of the bridge. An example of such an approach is the analysis presented in the work by Hong et al. (2011), where a framework for predicting the wind-excited response of a newly built suspension bridge was established by means of output-only identification and model updating. The identification of the bridge's natural frequencies and damping ratios was conducted using a modified version of the stochastic subspace identification, whereas the FE model updating was obtained by minimizing a weighted sum of the errors of the natural frequencies.

Dealing with datasets related to operational conditions involves handling data recorded over a long period of time, usually years. This allows engineers to analyze data recorded from the bridge when exposed to different environmental conditions, for example, summer versus winter. Different environmental conditions might alter the dynamic characteristics of a structure; for example, natural frequencies might vary up to 10% of their original value and this might be read, from an untrained eye, as a

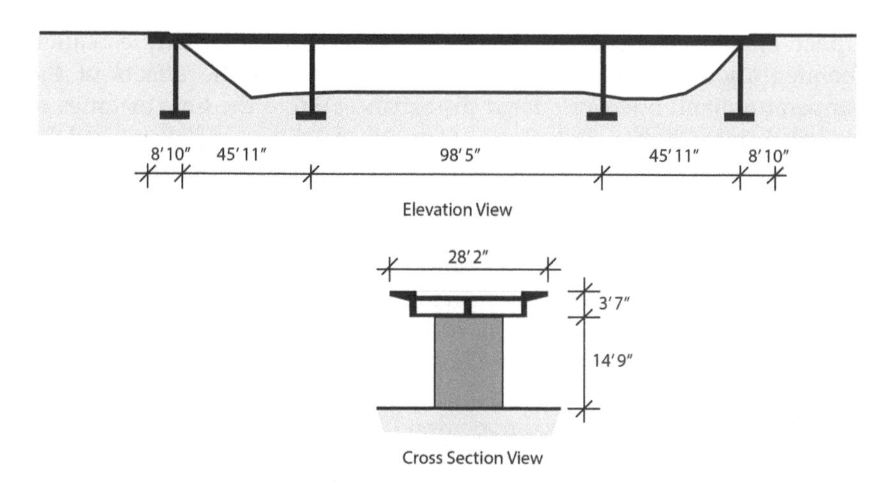

Figure 3-3. Schematics of the Z24 bridge.

consequence of the occurrence of damage in the structure. An example of such changes can be seen in the case of the identified natural frequencies of the Z24 bridge in Switzerland, a reinforced concrete bridge that was monitored for a year before being demolished (Figure 3-3). As for Tronci et al. (2019), Figure 3-4 shows the identified first and second natural frequencies of the bridge from the records spanning over 1 year of operations. The two frequencies present a substantial variation in their values in the 1,200 to 1,500 observation range: such a variation was found to be related to extreme cold temperatures that dramatically affected the stiffness of the pavement, inducing an increase in the structural frequencies.

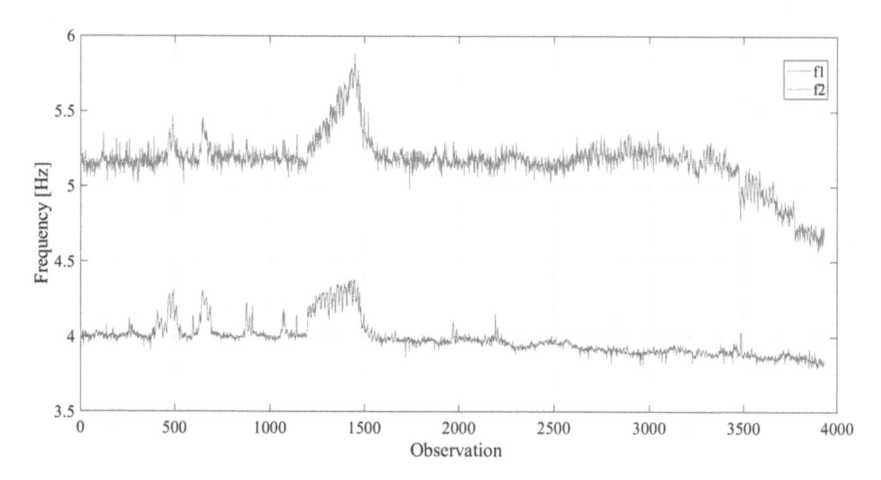

Figure 3-4. Identified first and second natural frequencies of the Z24 bridge (Switzerland).

Tronci et al. proposed an effective mathematical technique, called "cointegration" (Haichen et al. 2018), to eliminate the effects of the temperature (and other unrelated disturbances) from the time histories of the structural response. The effects of structural damage appear toward the end of the two graphs in Figure 3-4, where a progressive damage imposed on the bridge appears as a reduction of the identified natural frequencies.

Frequency domain methods rely on the use of some characteristic functions, for example, the frequency response function or the power spectral density function, that provide an indication of the frequency content of the recorded signal. By identifying peaks and valleys of such functions allows us to extract estimates of the natural frequencies, damping ratios, and mode shapes. Among these methods, one of the most effective is the frequency domain decomposition (Brincker et al. 2001), in which a singular value decomposition of the power spectral density function of the output signals is used to determine the dominant modes and the corresponding frequencies. One of the drawbacks of the frequency domain approach is the handling of closely spaced modes that require particular attention.

3.4 USE OF STRUCTURAL HEALTH MONITORING TO ENHANCE THE RECOVERY AND RESOURCEFULNESS OF HIGHWAY BRIDGES

3.4.1 Overview

The previous section discussed the use of SHM techniques to monitor two components of highway bridge resilience, specifically, robustness and redundancy. In this section, we will show how the use of such SHM techniques can be extended to enhance the other two remaining components of resilience according to NIAC (2009): recovery and resource utilization.

Let us start by observing that in a region that was subjected to a catastrophic natural (flood, earthquake, hurricane, and so on) or man-made (factory explosion, terrorist attack, and so on) event, response and recovery efforts heavily rely on this region's bridges and their connecting roadway network. The response and recovery resources (human, equipment, and so on) are usually located in different predesignated locations and need to be dispatched to the various sites in the shortest possible time. For example, the NYC Subway Action Plan (see MTA 2017) calls for 12 prepositioned Emergency Medical Technician teams throughout New York City for response during health emergency within the NYC Subway system. When response/recovery efforts are called for, during or after an event, the research/recovery resources will need to travel from one or more of the prepositioned locations to the locations that need those response/recovery resources. The transportation of goods and personnel from and/or to the sites of interest will be on the aforementioned bridge and roadway network (not counting air or waterway transports) and it will rely on the network's functionality. For efficient transport, the network will need to be in an adequate functional state; otherwise, response/recovery efforts might be delayed or even at risk.

Formally, the problem can be stated as a resource allocation problem on a transportation network. This is a well-studied problem; see the work by Luss (2012). Given a set of resource locations that is scattered on a bridge and roadway network, we need to find the optimal route from one, or more, of these locations to a given point on the network (the location that needs response/recovery help, e.g., the site of a flooding). In an emergency situation, the metric of interest is the travel time: A quick response to an emergency can result in fewer causalities in terms of human lives, better confinement of the damage, more effective use of resources, and a substantial reduction of cost. Hence, the goal is to optimize, in this case, minimize travel time (and consequently, cost) from one or multiple resource locations to the required point of interest. Unfortunately, there are several factors that can affect travel time (they can be considered as impedance factors). Among these impedance factors are travel distance and the state of repair of the bridges and roads along the potential travel routes. Minimizing travel distances within a network is a well-known graph network problem (Deo 1974). However, these theories rely on having a perfectly functional roadway network. Instead, minimizing travel distances during or immediately after a major event also requires providing information on the conditions of the network. Estimating, accurately, in real time, or near real time, the state of repair, during or after a disaster, of roads or bridges, either manually or using technological observations, is quite challenging. For example, during or immediately after a major event (e.g., major earthquake), accurate manual observations might not be available until after some time. It is in these circumstances where SHM processes can help in providing accurate estimates, which can enhance optimizing search and recovery travel times.

Here, using a simple bridge network, we offer two specific examples on how SHM efforts can help in such real/near real-time response/recovery efforts. In the first example, we assume that there is only a single search/recovery resource location, whereas the second example accounts for multiple response/recovery resource locations.

3.4.2 Example: Use of Structural Health Monitoring to Enhance Recovery with Single Resource Location (Nonredundant Resources)

As an example of how SHM can aid in recovery efforts, let us consider the simple network shown in Figure 3-5. The network contains two nodes:

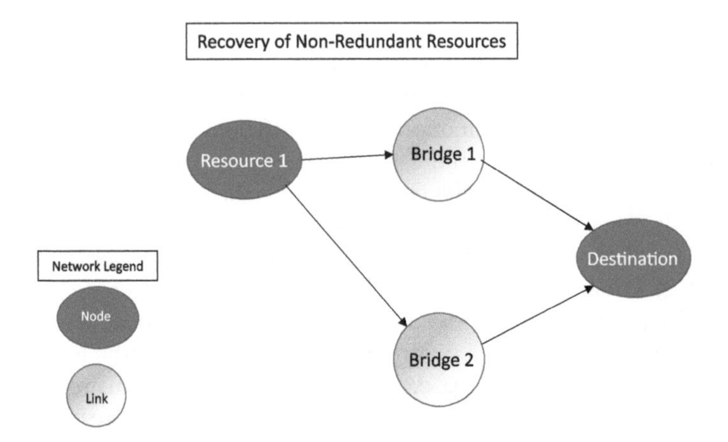

Figure 3-5. Simple recovery routing network.

Resource_1 and *Destination*. The two nodes are connected to two links: L_1 and L_2. The node *Resource_1* represents a location where needed resources during an abnormal event are located. These can be emergency resources such as personnel and equipment, fire, or police resources. The *Destination* node represents a location where help is needed during an abnormal event such as a failed infrastructure, fire, or health emergency. We assume that each of L_1 and L_2 contains (1) a set of roads with average travel times of $T_1 = 25$ min and $T_2 = 50$ min, respectively; and (2) a single bridge crossing on each link, *Bridge_1* and *Bridge_2*, respectively.

Furthermore, let us assume that, in normal conditions, the capacity of each bridge to carry emergency equipment, as monitored by the SHM systems previously described, is $Cap_1 = 100\%$ and $Cap_2 = 100\%$, respectively. Hence, both bridges, in their operational conditions, have a sufficient capacity to sustain the emergency equipment loading.

Pre-event emergency/recovery routes: The relative impedance (costs) of recovery routes along each of the links are, for *Bridge_1*,

$$C_1 = 25 \tag{3-1}$$

and, for *Bridge_2*,

$$C_2 = 50 \tag{3-2}$$

In this case, only the travel time will affect the link impedance. The robustness of the bridges along the routes will have no effect on the costs because both bridges are operating at full capacity (with regard to emergency equipment travel). Thus, an operational control center that has access to this information would easily determine that the optimal recovery route from *Resource_1* to *Destination* should be along L_1, because this would result in the optimal (minimum) cost.

During- and Post-event Emergency/Recovery Routes: An abnormal event affecting the region of the network might cause damage to one, or both, *Bridge_1* and/or *Bridge_2*, and such a damage might affect the optimal emergency/recovery routing. Let us assume that, from the analysis, in real time, of the data recorded by the SHM system during the event, it appears that *Bridge_1* has suffered some damage and its capacity has now been reduced to $Cap_1 = 75\%$, whereas *Bridge_2* has not suffered any damage ($Cap_2 = 100\%$). This implies that *Bridge_1* cannot carry emergency/recovery equipment safely, whereas *Bridge_2* is fully operational. Formally, the cost of traveling along L_1 now becomes

$$C_1 = \infty \tag{3-3}$$

Within the network, only *Bridge_2* still has enough robustness to carry emergency/recovery equipment in a safe manner. It is then obvious for the control center in charge of the routing, equipped in real time with such information, to decide that the optimal recovery route in this situation is along L_2.

3.4.3 Example: Structural Health Monitoring Role in Optimal Resource Management during Recovery (Some Redundancy in Resources Exists)

In the previous example, we have considered the case where there is only a single resource location in the network, and, as such, the network had no resource redundancy. Nonredundant resources can have a major negative effect on the resilience of such networks. For example, as in the previous case, if an abnormal event caused the capacities of the two bridges, as shown by an SHM process, to change such that $Cap_1 = 75\%$ and $Cap_2 = 90\%$, then the impedance of both the available routes become infinite ($C_1 = \infty$ and $C_2 = \infty$). This will render a resilience rating of the network almost nil, because there are no available recovery routes from the resource node to the destination node.

Let us now consider a more complex network where another resource node, *Resource_2*, is available. For simplicity, we assume that the two resource nodes are identical in capabilities. The network still has a single destination node, *destination*. Table 3-2 shows the different possible paths from the properties of the four possible combinations: resource nodes—bridge—the destination node. Before the event, it is clear that the optimal route is path $i = 1$: from *Resource_1* to *destination*. After the event, the value of SHM information and the added redundant resource node becomes obvious because the optimal path now, as revealed by the SHM process in real time, is path $i = 4$: from *Resource_2* to *destination* (Figure 3-6).

Table 3-2. Bridge Network Properties Before and After an Event.

Route ID, i	Path, L_i	Travel time, T_i (min)	Cap_i, Bridge capacity (%) according to the SHM process		C_i, Path relative impedance (cost)	
			Before event	After event	Before event	After event
1	*Resource_1–Bridge_1–destination*	25	100	75	25	∞
2	*Resource_1–Bridge_2–destination*	100	100	100	100	100
3	*Resource_2–Bridge_1–destination*	120	100	75	120	∞
4	*Resource_2–Bridge_2–destination*	70	100	100	70	70

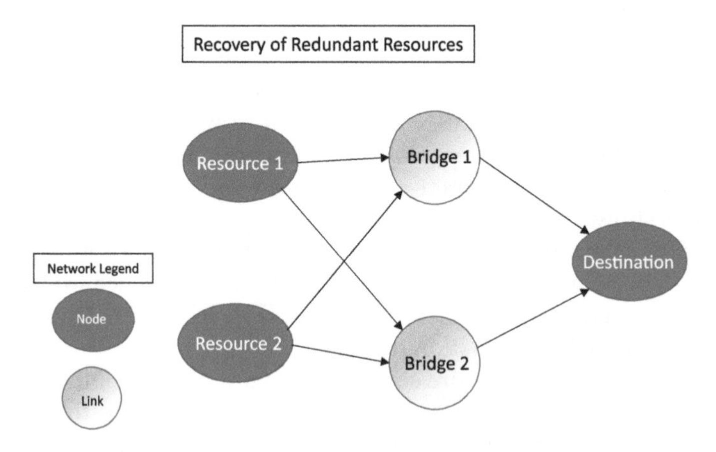

Figure 3-6. Redundancy and recovery routing network.

Cost–Benefit of SHM during Recovery: At this point, it would be important to conduct a cost–benefit analysis on the use of SHM to monitor the robustness of the bridge network along recovery routes. The SHM cost would be the actual monetary cost of building and maintaining the SHM system and infrastructure, whereas the benefits would be the recovery routing time savings that were realized by using SHM observations. For a general discussion on the cost–benefit of an SHM project, see Ettouney and Allampalli (2011), whereas, for a detailed process of computing SHM cost–benefit for the recovery phase of the resilience of infrastructure network, see Ettouney (2022).

3.5 BEST PRACTICES FOR MONITORING RESILIENCE OF HIGHWAY BRIDGES

In conclusion, it appears that it has become essential, in today's time, to take advantage of the advances in sensing technology in assessing the actual conditions of highway bridges for estimating the resilience of entire communities after major catastrophic events. Providing real-time information on the safety of bridges and on their capacity to carry different levels of loading will allow emergency operators to direct their resources in more effective (and economical) ways. The following best practices should be considered by engineers/owners/managers of roadway bridge networks:

1. Incorporate SHM systems and processes on critical bridges along major recovery/emergency routes. The cost of such systems is relatively low compared with the initial cost of a new bridge or with the cost of maintenance/rehabilitation programs. These systems must be reliable and redundant so as not to lose valuable information when needed, for example, for a sensor malfunctioning.
2. In a network of roadway bridges, it is important to monitor all the bridges that are along major transportation arteries. This will guarantee that, in the case of major catastrophic events, the structural conditions of the main evacuation routes are immediately assessed.
3. Incorporate processes that can identify, in real time, the capacities of bridges for at least three limit states (dead load, dead load + nominal live load, dead load + nominal live load + emergency/recovery equipment) as previously illustrated. These processes must provide reliable answers in real time, leaving the details to in-depth analyses that could be conducted at a later stage.
4. Perform a condition assessment of all the bridges in the network routinely. This operation will have multiple benefits: (1) it will help keep track of any modification occurring on a bridge because of reasons different from a major event (the effects linked to deterioration, e.g., gradual loss of prestressing, corrosion, fatigue, etc.), having always the latest model available; (2) it will allow engineers and owners learn about the structure, and so, to remove the environmental/operational effects; and (3) it will help with

long-term training of the engineers/managers in charge of emergency operations.

5. In the condition assessment process, include information from compatible databases. For roadway bridges, useful information can be found in the National Bridge Inventory database (FHWA 2017). These databases can complement the condition assessment process done using the field data and help with the results of the analysis.
6. Incorporate redundant resource sources as pertinent. Having resources allocated at different source points will offer multiple solutions that could eventually become the sole solution in the case of extended damaging events.

ACKNOWLEDGMENTS

The author would like to acknowledge the contribution of Dr. M. M. Ettouney for his guidance and suggestions on this manuscript.

REFERENCES

ASCE. 2017. "ASCE infrastructure report card." Accessed October 20, 2019/. www.infrastructurereportcard.org.

Brincker, R., L. Zhang, and P. Andersen. 2001. "Modal identification of output-only systems using frequency domain decomposition." *Smart Mater. Struct.* 10 (3): 441–445.

Deo, N. 1974. *Graph theory with applications to engineering & computer science.* Mineola, NY: Dover.

Ettouney, M. M. 2022. "Resilience management." Chapter 6 in *Objective resilience: Policies and strategies*, edited by M. M. Ettouney, MOP 146, 133–236. Reston, VA: ASCE.

Ettouney, M. M., and S. Alampalli. 2011. Vol. 1 in *Infrastructure health in civil engineering: Theory and components.* Boca Raton, FL: CRC Press.

FHWA (Federal Highway Administration). 2017. *National bridge inventory (NBI).* Washington, DC: FHWA.

Fraser, M., A. Elgamal, X. He, and J. P. Conte. 2010. "Sensor network for structural health monitoring of a highway bridge." *ASCE J. Comput. Civ. Eng.* 24 (1): 11–24.

Haichen, S., K. Worden, and E. J. Cross. 2018. "A regime-switching cointegration approach for removing environmental and operational variations in structural health monitoring." *Mech. Syst. Signal Process.* 103 (6): 381–397.

Hong, A. L., F. Ubertini, and R. Betti. 2011. "Wind analysis of a suspension bridge: Identification and FEM simulations." *J. Struct. Eng.* 137 (1): 133–142.

Hou, R., S. Jeong, J. P. Lynch, and K. H. Law. 2020. "Cyber-physical system architecture for automating the mapping of truck loads to bridge behavior using computer vision in connected highway corridors." *Transp. Res. Part C* 111: 547–571.

Linderman, L. 2019. *Displacement monitoring of I35W Saint Anthony falls bridge with current vibration-based system.* Final Rep. Saint Paul, MN: Minnesota Dept. of Transportation.

Lus, H., R. Betti, and R. W. Longman. 2002. "Obtaining refined first order predictive models of linear structural systems." *Earthquake Eng. Struct. Dyn.* 31 (7): 1413–1440.

Luss, N. 2012. *Equitable resource allocation: Models, algorithms, and applications.* Hoboken, NJ: Wiley.

Manyena, S. B. 2006. "The concept of resilience revisited." *Disaster* 30: 434–450.

Mosquera, V., A. W. Smyth, and R. Betti. 2012. "Rapid evaluation and damage assessment of instrumented highway bridges." *J. Earthquake Eng. Struct. Dyn.* 41 (4): 755–774.

MTA (Metropolitan Transit Authority). 2017. *NYC subway action plan.* New York: MTA.

NIAC (National Infrastructure Advisory Council). 2009. *Critical infrastructure resilience: Final report and recommendations.* Washington, DC: NIAC.

NSC (National Security Council). 2011. "Presidential Policy Directive/ PPD-8: National preparedness." Accessed May 26, 2018. https://www.dhs.gov/presidential-policy-directive-8-national-preparedness.

Tronci, E. M., R. Betti, M. De Angelis, and M. Q. Feng. 2019. "Environmental effects on cepstral coefficients and their removal for structural performance assessment." In *Proc., 9th Int. Conf. on Structural Health Monitoring of Intelligent Infrastructure, ISHMII, St. Louis, Missouri.* www.ishmii.org.

CHAPTER 4

OBJECTIVE RESILIENCE MONITORING FOR RAILROAD SYSTEMS

Katherine A. Flanigan, Marlon Aguero, Roya Nasimi, Fernando Moreu, Jerome P. Lynch, Mohammed Ettouney

4.1 INTRODUCTION

4.1.1 Importance of Resilient Railroad Systems

Railroads (RRs) are a vital transportation system on which the national economy is highly dependent. For example, it is estimated that the seven largest RRs in the United States (the so-called Class I RRs) currently move in excess of 1.2 million ton miles (1.75 million metric ton kilometers) of freight per year (DOT 2015). Furthermore, 40% of the nation's freight tonnage is carried on rail (AAR 2015). Considering current trends, North American RRs are projected to exceed their throughput capacities at many network locations over the next 20 years, resulting in an urgent need for advance planning of infrastructure expansion (FHWA 2006). Included in the rail sector is passenger rail, which services the mobility of people. In the United States, Amtrak supports long-distance rail travel with more than 300 trains traveling over 22,000 mi (35,406 km) of routes daily; there is an additional 29 light, 25 commuter, and 15 heavy rail agencies (Baylis et al. 2015). Given the critical role that RRs (freight and passenger) play in the economy, there is an urgent need to ensure that they are resilient against hazards that can disrupt their operations. Railroads are regulated by multiple federal agencies, including US Department of Transportation (DOT) units such as the Federal Railroad Administration (FRA), the Federal Transit Authority (FTA), and the Surface Transportation Board. The Department of Homeland Security (DHS) is another key federal agency that offers security and resilience oversight through the Transportation Security Administration (TSA).

Investing in a resilience strategy is a vital step toward ensuring the continuity of RR operations when hazardous events occur. Major North American freight RRs have been strategic in their investments, aiming to ensure the resilience of their assets and networks (Baylis et al. 2015). These investments are justified because they help RRs manage their risks and ensure long-term financial stability for the companies and their investors. Public-owned rail systems, such as commuter rail, also prepare for disruptive events, but their investments in resilience preparation often lag private RRs as other needs often demand resources ahead of risk management.

4.1.1.1 Resilience Definition. A key step in managing resilience is being able to fully quantify what it means to be resilient. The National Infrastructure Advisory Council (NIAC) defines infrastructure resilience as the ability to anticipate, absorb, adapt to, and rapidly recover from damage or stressors to reduce the duration and magnitude of diminished operational levels (NIAC 2009). The time-based trajectory of operational level (Figure 4-1) was first introduced by Haynes (M. E. Haynes, personal communications, 2001) and then formalized by Bruneau et al. (2003). It depicts performance as a function of time, $Q(t)$, and has been used to describe the nature of asset (or community) resilience before, during, and after a disruptive event. Figure 4-1 applies conceptually regardless of whether the "asset" under consideration refers to an infrastructure component (e.g., bridge), a network system (e.g., an RR network inclusive of track, bridges, stabilized slopes), or even larger communities. The first portion of the functional resilience curve (Figure 4-1) represents the normal operational level of an infrastructure asset. During this period, the asset operates under its expected load profile and behaves as intended. If at some point in time a disruptive event occurs, the asset likely responds with some degradation in its performance, α. The change in the operational level of the asset is relatively instantaneous (although it need not be and

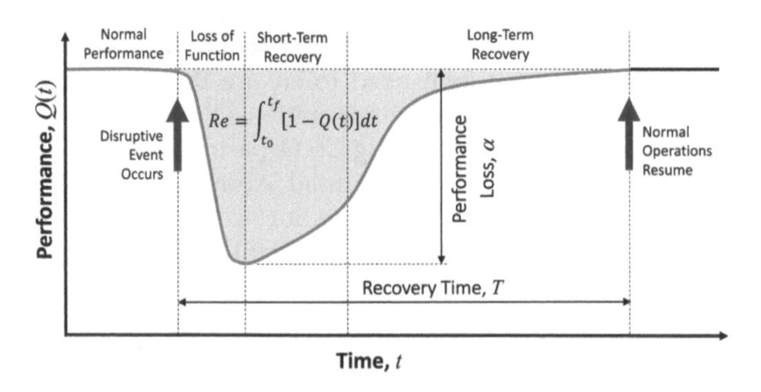

Figure 4-1. Classical functional resilience curve.

can occur over some time frame, as shown in Figure 4-1) and represents a fundamental change in the system, often in the form of damage to the system. Figure 4-1 illustrates that maximizing asset resilience is equivalent to minimizing the area $\mathrm{Re} = f(Q,t)$ which is the area between the diminished operational level and the normal operating level starting when the disruptive event occurs and ending when normal operations resume. This is achieved by decreasing the initial reduction in performance, α, decreasing the recovery time, T, and increasing the recovery rate. Gerasimidis and Ettouney (2022) observe that all popular resilience definitions exclusively include these two variables (i.e., performance level and recovery time) and, as a result, call this high-level definition the "universal" definition of resilience.

Beyond the universal resilience components, many authors have identified second-tier components that control either performance, Q, or time to recovery, T. NIAC (2009) introduced a set of subcomponents that can be used to quantify resilience. NIAC identifies four subcomponents: robustness, redundancy, resourcefulness, and recovery. These four components, herein referred to as the 4Rs, can be evaluated objectively and are related to the two universal resilience components, Q and T, where $R_i = f_i(Q,T),\, i = 1,2,\ldots,4$. The *robustness* ($R1$) of the system corresponds to how well designed the asset is to withstand a disruptive event; a more robust asset will exhibit a smaller reduction in the operational level than a less robust asset. *Redundancy* ($R2$) refers to internal functional elements that can be engaged when primary functional elements are lost or degraded. Redundancy can strongly influence the magnitude of operational-level reduction during a disruption. Immediately following a disruptive event, the asset owner/manager will devote resources to execute a recovery strategy for the asset. This *resourcefulness* ($R3$) has two primary aims: (1) to speed up the recovery process as quickly as possible, and (2) return the asset to its full operational level of performance. The asset owner/manager controls the resource flows that can be devoted to the asset recovery, thereby controlling the slope of the operational-level curve after the event. Similarly, *recovery* ($R4$) is the capacity of an organization to execute its recovery efforts that control the response time. Recovery includes having alternate options for addressing the loss of the operational level of the asset. This can, in turn, strongly influence the shape of the functional resilience trajectory. A single estimate for resilience can be estimated as a function of these subcomponents and is defined as $\mathrm{Re} = f(R1, R2, R3, R4)$. These quantifiable and objective properties of resilience strongly influence the recovery parameters illustrated in Figure 4-1 (i.e., the shape of the functional resilience trajectory after a disruptive event occurs).

4.1.1.2 Resilience Management of Railroad Systems. The resilience of any system, including RR systems, is a complex paradigm; attaining *optimal* resilience is an even more complex endeavor. Consequently, a comprehensive resilience management approach is essential to achieve optimality (Ettouney 2014, 2022). Resilience management (RM) is composed of the following five components:

1. Resilience assessment: assessing the current state of the RR system's resilience, Re;
2. Resilience acceptance: making a decision as to whether Re, as obtained in Step 1, is acceptable;
3. Resilience enhancement: if Re is *not acceptable*, determining an optimal way to enhance the resilience to return it to an acceptable threshold;
4. Resilience monitoring (RMo): realizing that as the state of the RR system resilience degrades over time (especially as the demands imposed by hazards on the RR system increase), it is essential to monitor Re as time passes and repeat Steps 1 through 3, as necessary; and
5. Resilience communications: communicating Steps 1 through 4 to the RR system's different stakeholders is paramount for the successful implementation of resilience management efforts.

Figure 4-2 illustrates the different components of RM and places an emphasis on the resilience monitoring cycle within the RM system because

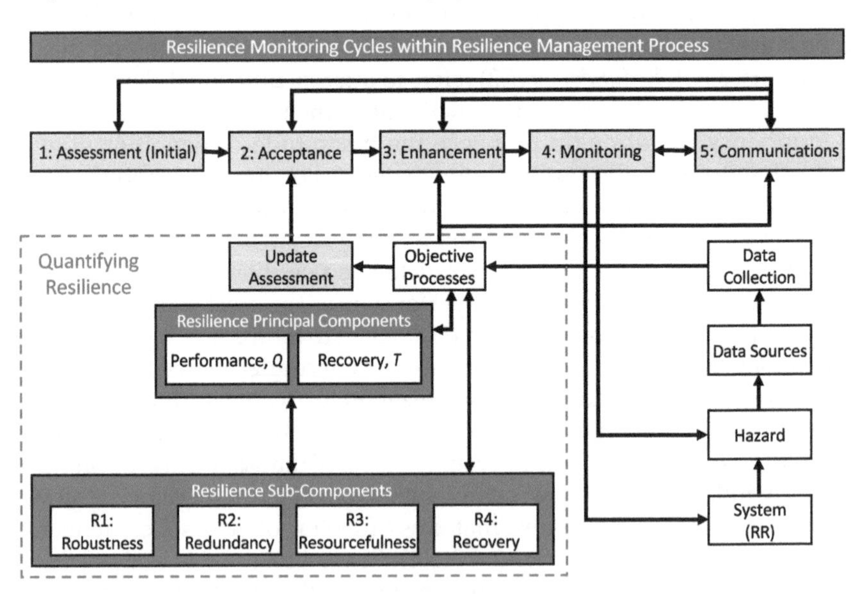

Figure 4-2. Resilience monitoring cycle as related to the resilience management system. Resilience principal components are found in all relevant resilience definitions, while resilience subcomponents might vary depending on the governing resilience definition.
Source: Gerasimidis and Ettouney (2022).

this is the main focus of the chapter. It also illustrates the different aspects of resilience monitoring and how collecting data from pertinent data sources, which are at the heart of any monitoring system, can be utilized to (1) evaluate the 4Rs; (2) evaluate the two universal resilience components, Q, and T; and (3) update the system resilience to complete the management cycle of Steps 1 through 5 as described.

4.1.1.3 About This Chapter. As previously stated, this chapter of the Manual of Practice (MOP) explores different issues related to the resilience monitoring of RR systems. It begins with a description of the profiles of some of the most important hazards considered by RR managers and owners. Specifically, the various stressors that pose a risk to asset and network operations are presented. Second, many of the available data sources that can be used to objectively quantify the components of resilience are described in detail. Third, the NIAC's four resilience components are described in greater detail in the context of the RR sector. A number of case studies follow to illustrate the quantification of resilience in various rail applications.

Observing that monitoring robustness has, in general, been the primary aim of the structural health monitoring (SHM) and nondestructive testing (NDT) fields (Ettouney and Alampalli 2011a, b; Lynch 2007; Lynch et al. 2016; Flanigan et al. 2020b), the authors posit that traditional SHM and NDT constitute a subset of resilience monitoring (i.e., comprehensive and optimal resilience monitoring requires more than just SHM and NDT). This observation is leveraged near the end of the chapter when the principle of resilience monitoring (PRM) and three related lemmas are presented. The PRM and associated lemmas aim to project a general basis for performing comprehensive and ultimately rewarding resilience monitoring practices as part of a resilience management program for civil infrastructure systems (with an emphasis placed on RR systems in this chapter). The chapter ends with the presentation of pertinent conclusions and a set of recommended practices.

4.2 HAZARD PROFILE OF RAILROADS

Freight RRs have unique risk profiles compared with other transportation and civil infrastructure systems because they support regional supply chains and are managed by for-profit corporations. With little ability to reroute rail traffic in the face of network closures, it is essential that RRs understand the wide range of disruptions that can lead to diminished operational levels. These disruptions include, but are not limited to, natural hazards, asset failures, collisions/derailments, and terrorism. Each of these disruptive events will be described in greater detail.

4.2.1 Natural Hazards

Depending on their geographic location, RRs are exposed to a number of natural hazards, including seismic activity, tropical storms, fires, and other extreme environmental conditions that can interrupt operations. RRs have historically been robust in the face of seismic disasters because of preparation for such events (Moreu and LaFave 2012). In recent years, flooding and hurricanes have caused significant damage to RR operations, requiring considerable efforts to resume operations after disasters (ASCE 2006). The interruptions experienced during recent floods and hurricanes suggest that the rail industry is less resilient to these hazards, especially when considering the prediction that storm events will likely increase in intensity and frequency under climate change. Fires also pose great threats to RRs because RR tracks are often supported by wood ties (although alternate materials such as concrete are now being used for ties) and a large number of timber bridges are still in service. For example, 25% of the national rail bridge inventory is constructed out of timber (Moreu et al. 2014). This relatively high percentage does not translate directly to a similarly high percentage of use because most wood RR bridges are located in short lines and branch lines with lower traffic density. While RRs treat timber infrastructure with chemicals to protect them against destructive pests (e.g., termites), deterioration from pests is also possible.

4.2.2 Aging Inventories

Aging is in many respects a hazard as it poses a serious risk to safe and efficient RR operations. Railroad owners and managers protect their infrastructure inventories against aging by acknowledging that deterioration of structural properties by fatigue and corrosion is expected. Natural deterioration (e.g., corrosion) can be accelerated depending on the location of the infrastructure asset because of exposure to the elements. To manage aging risks, RRs make extensive use of visual inspection methods followed by proper maintenance; these methods slow the detrimental effects of natural aging and deterioration.

4.2.3 Collisions and Derailments

Recent derailments in the United States illustrate the importance of taking proactive preventative actions against factors causing these events. For example, on July 18, 2013, a northbound CSX Transportation train derailed on the Metro-North Railroad in New York because of excessive track gauge (NTSB 2013). Derailments can cause major disruptions in the operations of an RR system (Garg et al. 2019). While derailments have decreased significantly over the last few decades, railroads still prioritize reducing derailments with proactive prevention and research (Liu et al. 2017). Derailments are caused by either human errors or infrastructure deficiencies. Human-related derailments can be avoided by improving the training practices of RR personnel. Infrastructure deficiencies that might cause derailments can be prevented through proactive track maintenance, repair, and replacement.

Vehicular collisions present another serious hazard. For example, damage can occur when trucks collide with elevated tracks or ships strike bridge piers. Such collisions are a serious issue because they can result in bridge damage and undetected distortions to the track. A serious issue arises when these collisions occur and are not reported. When collisions are not reported, the safety of the bridge and track cannot be assessed. For example, a barge collided with the Big Bayou Canot Bridge in Alabama on September 22, 1993, resulting in permanent displacement of the track; an Amtrak train on the displaced track derailed, leading to 47 fatalities shortly after, and 103 passengers severely injured (NTSB 1993). On-track obstructions such as vegetation, wildlife, and man-made obstructions (e.g., nonrail vehicles) can also cause collisions and derailments. For example, collisions between locomotives and vehicles at grade crossings commonly occur and are a leading cause of railroad-related deaths (GAO 2018).

4.2.4 Terrorism

RRs must observe and prevent acts of terrorism as they can be extremely destructive to both the RR and communities surrounding RR operations. From 1998 to 2003, there were 181 incidents of terrorism on rail infrastructure, resulting in 431 deaths globally (Riley 2004). Particularly dangerous freight (e.g., those transporting explosive chemicals or nuclear materials) can serve as a highly attractive target to those individuals and groups seeking to intentionally release lethal fumes and radiological materials into the environment, especially in dense urban areas. Since September 11, 2001, DHS and RR companies have established open lines of communication that can be enacted immediately after a disaster (Riley 2004). To increase preparedness, RR companies train their staff to be alert for any sign of terrorist activity near their infrastructure (Strandh 2017). Security systems are also implemented near critical RR infrastructure to alert bridge officials of any unauthorized trespassing.

4.3 DATA SOURCES TO MONITOR AND INCREASE RAILROAD OBJECTIVE RESILIENCE

Measurements, data, and information are necessary to quantify (or "objectify") resilience. This section aims to illustrate the diversity of available data sources and sensing technologies that can serve as inputs to the calculation of objective metrics that quantify the different components of RR resilience.

There are several diverse data sources that can be leveraged to help quantify the 4Rs within an objective framework for assessing the resilience, Re, of RR networks. These data sources are roughly (but not exclusively) sorted into five categories. First, automated sensing technologies serve as a means of monitoring the performance of network assets including bridges, culverts, and geotechnical systems (e.g., piers, retaining walls, embankments) over long observation periods such as years and decades. Such data offer measurement of component responses to both environmental and operational conditions. Processing of the data can inform a holistic view of the capacity and demand of operational infrastructure in RR networks. Second, wayside detection methods help to identify damage to rolling stock such as hot bearings, sliding wheels, unexpected wheel impacts, and cracked wheels. Wayside detection methods are widely employed by RRs as a means of ensuring safe operations. These methods identify wheel- and axle-bearing defects that lead to excessive wear on tracks and can lead to train derailments. Some of these tools also offer measurement of axle loads imposed by locomotives and rail cars on track infrastructure including bridges. Third, global positioning systems (GPS) provide spatial and temporal information to inform decision makers of train locations (along with their freight), aid in train control, help to avoid train-to-train collisions, and offer data for tracking the rolling live load throughout the network. Fourth, manifest data contain detailed information about the weightage and contents of cargo [e.g., toxic inhalation hazard (TIH) materials], which is critical to assessing the risks of component and network failure. Fifth, vehicle-based (on-board) data sources are also available, including from track geometry vehicles, rail flaw detection vehicles, ground-penetrating radar (GPR), vehicle track interaction (VTI)–equipped locomotives, on-board machine vision and light detection and ranging (LiDAR), gauge-restraint measurement system (GRMS) vehicles, and by-rail inspections. These five data sources, including a number of available technologies, corresponding types of measurements, sensing objectives, and resilience components that benefit from these types of information, are detailed in Table 4-1.

In addition to the five data sources presented in Table 4-1, there are other sources of data pertinent to resilience monitoring, including weather monitoring networks/services and earthquake notification networks. These systems produce data that are used to enhance RR preparedness, resource utilization, and recovery operations. These types of hazard monitoring stations cover the entire network for the subscribing RR.

Table 4-1. Data Sources Commonly Used for Rail Infrastructure Management.

Data type	Available technologies	Type of measurement	Objective(s)	Resilience metric that might utilize the resulting information
Automated sensing technology	Triaxial accelerometers (Flanigan et al. 2020a)	Acceleration along three orthogonal axes	Measures asset response (vibrations) to loads; can be used for finite-element model updating via modal parameters	• Performance, *Q* • Time to recovery, *T*, *R*3 • Robustness, *R*1 • Resourcefulness, *R*2 • Redundancy, R4
	Uniaxial accelerometers (Flanigan et al. 2020a)	Acceleration along a single axis	Measures asset response (vibrations) to loads; used to measure local component accelerations (which can be used to assess axial loads on components)	• Performance, *Q* • Time to recovery, *T*, *R*3 • Robustness, *R*1 • Resourcefulness, *R*2 • Redundancy, *R*4
	Strain gauges (Flanigan et al. 2020a)	Strain	Measures strain response (dynamic and static) of structural components; used to assess fatigue in metallic components	• Performance, *Q* • Time to recovery, *T*, *R*3 • Robustness, *R*1 • Resourcefulness, *R*2 • Redundancy, *R*4
	LVDTs (Liu et al. 2018)	Linear position	Measures relative displacement in structural systems; can be used to measure structural displacement relative to foundation	• Performance, *Q* • Time to recovery, *T*, *R*3 • Robustness, *R*1

(*Continued*)

Table 4-1. (*Continued*)

Data type	Available technologies	Type of measurement	Objective(s)	Resilience metric that might utilize the resulting information
	Piezoelectric sensors (Lynch et al. 2017)	Vertical tactile force	Offers means of measuring tactile action; can be used to detect axles on tracks and bridges	• Performance, Q
	Geophones (Flanigan et al. 2020a)	Vertical velocity	Measures vibrations in geotechnical (or structural) systems; can be used to trigger other monitoring systems by detecting arrival of trains	• Performance, Q • Time to recovery, T, $R3$ • Robustness, $R1$
	Thermistors (Lynch et al. 2017)	Temperature	Effects of temperature on structural elements	• Robustness, $R1$
	Wired data acquisition systems (Ngamkhanong et al. 2018)	—	Mainstay approach to collecting data from assets using coaxial wiring between sensors and data logger	• Performance, Q • Time to recovery, T, $R3$ • Robustness, $R1$ • Resourcefulness, $R2$ • Redundancy, $R4$
	Wireless sensing nodes (Swartz et al. 2005)	—	Collect, process, and wirelessly transmit data	• Performance, Q • Time to recovery, T, $R3$ • Robustness, $R1$ • Resourcefulness, $R2$ • Redundancy, $R4$

Wayside detection (internal safety operations manage-ment)	Wheel impact load detection (WILD) (L.B. Foster Company 2019, Barke and Chiu 2005)	Strain	Measure load demand in the form of axle weights; identify out-of-round wheels	• Robustness, *R*1
	Hot-bearing "hotbox" detectors (UP 2017)	Temperature (through infrared radiation)	Identify faulty bearings to prevent seizure of bearings on the track	• Robustness, *R*1
	Ultrasonic wheel-defect detection (UP 2017)	Ultrasonic	Detect cracks in wheels	• Robustness, *R*1
	Acoustic bearing detectors (UP 2017)	Acoustic emissions (using microphone)	Identify bearing defects such as spalling of the cup or cone, and seamed, spalled, or etched rollers	• Robustness, *R*1
	Wheel profile detectors (UP 2017, Barke and Chiu 2005)	Digital image acquisition	Identify defects impacting flange thickness, flange height, diameter, rim thickness, and angle of attack, by reconstructing the exact profile of the wheel tread; store images to determine wheel wear rates and remaining life	• Robustness, *R*1

(*Continued*)

Table 4-1. (*Continued*)

Data type	Available technologies	Type of measurement	Objective(s)	Resilience metric that might utilize the resulting information
Global positioning system (GPS)	GPS and GPS-based positive train control (PTC) (National Coordination Office 2018)	Geolocation (spatial mapping) and time information	Offers real-time information on train locations often visualized in RR command centers; used in train control to prevent derailments owing to high train speeds, train-to-train collisions, unauthorized entry into work zones, and the movement of trains through misaligned track switches; can be used (in tandem with manifest data) to track live load in networks	• Performance, Q • Time to recovery, T, $R3$ • Robustness, $R1$ • Resourcefulness, $R2$ • Redundancy, $R4$
Manifest data	Automated commercial environment (ACE) (USCBP 2019)	—	Collect detailed information about cargo, including Customs Bill ID, Equipment ID, border crossing, shipping quantity and unit of measure, weight, among other critical information to assess freight weight and value	• Performance, Q • Time to recovery, T, $R3$ • Robustness, $R1$ • Resourcefulness, $R2$ • Redundancy, $R4$
Onboard monitoring systems	Track geometry vehicles (Li et al. 2017)	Track geometry	Collect information of the track alignment while on the track	• Performance, Q • Redundancy, $R4$ • Robustness, $R1$

Rail flaw detection vehicles (Magel et al. 2016)	Rail head deterioration	Use ultrasonic equipment to detect flaws in the rail	• Performance, Q • Redundancy, $R4$ • Robustness, $R1$
GPR, VIT equipped locomotives (Xu et al. 2018)	Defects on profile or subgrade	Ground-penetrating radar	• Performance, Q • Robustness, $R1$ • Resourcefulness, $R2$ • Redundancy, $R4$
On-board machine vision and lasers (Rakoczy et al. 2016)	Gauge, alignment, and track surface	Installed in special vehicles and cars equipped specially for monitoring	• Performance, Q • Redundancy, $R4$ • Robustness, $R1$
LiDAR machine vision (Lu et al. 2019)	Modulus estimation	Installed in car with limitations of speed under 40 MPH	• Performance, Q • Robustness, $R1$ • Resourcefulness, $R2$ • Redundancy, $R4$
GRMS vehicles (McHenry and LoPresti 2016)	Gauge-restraint measurement system	Testing the ability of the track to resist gauge widening	• Performance, Q • Robustness, $R1$ • Resourcefulness, $R2$ • Redundancy, $R4$
Hi-rail vehicles (Falamarzi et al. 2019)	NDT sensors, cameras, GPS, gyroscopes	Traditional on-rail car that can transition between road and rail with track inspectors. Enable human inspection and assessment in the field as well as sensors	• Performance, Q • Time to recovery, T, $R3$ • Robustness, $R1$ • Resourcefulness, $R2$ • Redundancy, $R4$

4.4 4RS FOR RAILROAD INFRASTRUCTURE

The following subsections summarize the main metrics of resilience (i.e., the two principal components, Q and T, as well as their subcomponents, the 4Rs) in relation to RR systems and illustrate a direct and explicit mapping between available data sources and the 4Rs (Figure 4-3). This section will reveal that the principal components of resilience and their subcomponents are inherently interdependent. In addition, data sources available in the RR industry can be used as inputs to quantify multiple resilience components. For example, Table 4-1 already illustrates that all of the data sources can aid in all, or most, of the resilience metrics. Characteristics of the resilience components that are unique to RR infrastructure (as opposed to other infrastructure classes such as pipelines and highways) are emphasized in this section. The descriptions provided are not intended to be exhaustive but rather to represent the current practices used by RRs to enhance their resilience. The section is specific to freight RRs, but many of the resilience practices presented herein are universal and would apply to commuter RRs as well.

4.4.1 R1: Robustness

The robustness of RR infrastructure can be measured by the ability of a network or component in the network to maintain its capacity after a

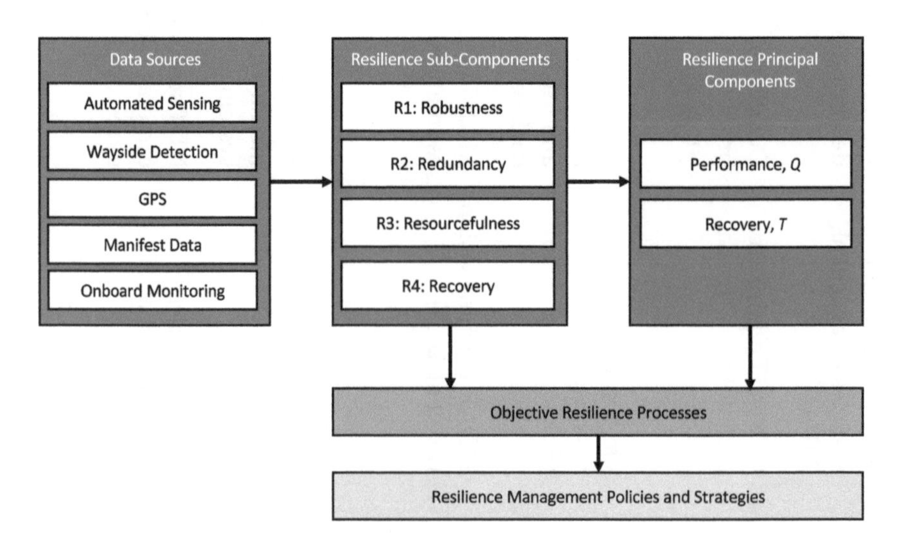

Figure 4-3. Flow of data from their sources through the components of objective resilience to ultimately aid in executing resilience management policies and strategies.

disruptive event. The capacity of RR infrastructure is defined in terms of maintaining the rating of individual infrastructure (e.g., rated E-Cooper loads) or the number of millions of gross tons (MGT) that are carried by the asset over a given period of time. Manifest data and wayside detection methods provide detailed information about cargo contents and high-resolution axle load measurements of trains moving through a network, respectively. These data sets quantitatively inform the number of MGT carried over an asset or within a network during a fixed period of time.

Track condition is a metric of objective resilience that is unique to RRs (as compared to other types of infrastructure systems). Currently, RRs quantify the condition of their track systems by measuring track geometries across time and space according to FRA requirements. Under FRA guidelines, different track considerations need to be considered to quantify track condition. This includes, for example, distinguishing between tangent or curved tracks, or portions of tracks with slopes (FRA 2010, 2016). Special track considerations are given to track locations within a network that are prone to accelerated deterioration in addition to sections of tracks that include sidings, single and double tracks, RR yards, frogs, switches, and diamonds. Rails are routinely inspected by human inspectors either walking on the track or observing the track from a slowly moving vehicle (CFR 2018a). Inspectors are seeking out a wide range of track defects that could pose a risk to safe operation of the track. In addition, nondestructive inspection (NDI) technologies including ultrasonic methods have been used to offer a more quantitative approach to assessing track conditions. Wayside detection methods are used to quantify the robustness of in-service tracks and the condition of the rolling stock traveling over them (such as train wheels). These methods, outlined in Table 4-1, are widely employed by RRs as a means of ensuring safe operations by identifying damage in rolling stock (e.g., hot bearings, sliding wheels, unexpected wheel impact, cracked wheels) that can lead to track damage and train derailments.

The safe condition of RR bridges is also critical to maintaining network robustness. As opposed to highway networks, which have approximately 1 bridge per 10 mi (16.1 km) of road, RR networks include approximately 1 bridge per 1.4 mi (2.25 km) of track (Moreu et al. 2017a). This high bridge-to-track mile ratio is because of the need for RRs to maintain grade across their entire network. Consequently, RRs must pay special attention to the condition of their bridges. In general, RR bridges are constructed from three types of materials: steel, concrete (including masonry), and timber. Historically, RR bridges have been classified into 14 types, depending on their design and functionality; however, structural systems comprising simply supported spans have been the preferred design strategy for RR bridges. This is because simply supported spans can be easily repaired and replaced as compared to multispan or continuous structures that are common in highway networks. More recent considerations in RR bridge design include the use of new materials and new loading conditions to accommodate evolving demand requirements such as higher freight loads and train speeds. To ensure that bridges remain in good condition and are safe for operation, the FRA requires (at a minimum) annual visual bridge inspections (CFR 2018b). However, many RRs adopt more frequent inspection procedures for some of their bridges, including quarterly inspection. While visual inspection is the dominant inspection approach, new sensing and NDI technologies are beginning to be adopted to offer more quantitative evidence of bridge condition and behavior (Moreu et al. 2017a).

Automated sensing technologies can be used to monitor the long-term robustness of assets subject to both environmental and operational conditions. Compared with standard visual inspection practices, data collected from automated sensing technologies can be used to quantify bridge conditions, asset ratings, and compliance with serviceability limits. Fundamentally, measurement data offer an objective measurement of component capacities and load demands. Reliability methods serve as one objective analytical framework that can be used to assess the safety of an asset with respect to its capacity and demand. Using data generated from automated sensing technologies as inputs to inform the capacity and demand within a reliability framework, the level of safety of an asset can be quantified with respect to a failure limit state (Nikolaidis et al. 2004), where a limit state can be governed by either single component or system-wide failure mechanisms. The goal of reliability methods is to calculate the reliability index, β, which is directly related to the probability of failure and an objective measure of safety.

The robustness of RR operations also requires that special attention be paid to proper design practices, maintenance, and replacement of soils and geostructural components such as tunnels, banks, fills, embankments, and earth structures. To simplify designs, maintenance, and replacement efforts, RRs have elected to replace old RR bridges with culverts where possible in recent years because they are more robust in the face of disruptive events.

In addition, the performance of RR infrastructure assets that control network operations, such as switchyards, intermodal facilities, and operational control headquarters, contributes to RR robustness. These individual assets control the system-level RR operations and are responsible for network signals including the implementation of positive train control (PTC). PTC is already being enforced in the United States to increase RR robustness by reducing the negative impacts of human errors on RR operations (DOT 2018).

Beyond the condition of RR physical assets (e.g., tracks, bridges, embankments, signaling systems), RRs have also developed means of tracking locomotives and rail cars in near real time in their systems using GPS transponders. This allows RRs to have real-time situational awareness during accidents and disasters. Railroads also perform rail threat assessments allowing them to identify vulnerabilities to identified threats (namely, terrorism) and to address them to ensure operational robustness. Often, measures to address vulnerabilities include increased security of RR assets and better management of rail cars carrying TIH materials (GAO 2009).

4.4.2 R2: Redundancy

To ensure that operations continue in the event of disruptions, RRs introduce redundancy to provide excess capacity at the component and network levels. RRs facing disruptive events reduce the risks of network capacity reductions by collaborating with other RR companies and government organizations (e.g., DOT, DHS, military) to safely share tracks and operational resources during emergency situations (AAR 2019a). This is enabled by objective data sources such as GPS and manifest data that enable RRs to reroute trains when disruptions in a network occur. Rerouting is done either by moving trains onto other network lines within their own networks or by using lines operated by different companies.

Even within the network itself, RRs will consider component redundancy. For example, switch points are vital to allowing RRs to reroute rail traffic as needed. Given that switches are single points of failure that could prevent the rerouting of trains, RRs are beginning to explore redundancy in switch designs to ensure proper operation at all times (Bemment et al. 2016). There is also component redundancy in the design of the communication networks on which RRs rely both for operations and for the observation of operational performance in real time.

At the individual structural level, RRs understand the consequences of failed components on the overall safety of their structures. In some cases, redundancy is not present in the design of the structural system. For example, truss bridges used in rail networks are often without component redundancy, which could result in bridge failures when a single structural element fails. Automated sensing technologies can be used to quantify the safety of high-priority and nonredundant assets and components; this information can be used to inform and prioritize maintenance to reduce the likelihood of failure by considering where redundancy does or does not exist in the bridge system.

4.4.3 R3: Resourcefulness

The resourcefulness of RRs needs to be managed to ensure that resource flows align with optimal decision-making strategies after a disruptive event. Resourcefulness is described as the ability of an RR to recover operations, including the ability to reroute traffic and maintain operations even when an asset is disrupted or while it is being repaired.

Training is a critical resource that is integral to the recovery of RR operations after a disruption. Railroads, especially the larger RR companies, extensively train their operational personnel and higher-level management in safety and emergency response. Training articulates how the RR and its personnel should respond to disasters. Incident response often requires decisions to be made as quickly as possible. Incident response plans and rehearsal of these plans under different scenarios constitute one approach that RRs adopt to maximize their capacity to respond to disruptions (CN 2019). Integral to response plans is the relocation of supplies, personnel, and rail traffic prior to disasters that can be forecasted (e.g., tropical storms). This allows for supplies and personnel to be rapidly (re)deployed immediately after the event. Included in the response of RRs is the deployment of personnel (in-house or contractors) who can inspect critical structures and other RR infrastructure. As previously mentioned, these inspections are largely vision-based and are used to provide a quick assessment of conditions. The emerging automated sensing technologies, wayside detection, and onboard sensing methods listed in Table 4-1 can be used to augment or replace visual inspection and quantify the safety and performance of an asset. An objective knowledge of the safety of damaged assets across a network enables owner/managers to introduce a quantitative notion of priority, thereby guiding the allocation of resources.

Railroad responses (including those scripted by incident response plans) are designed from RR command centers that always have critical decision makers on duty who can react to real-time conditions informed by data from the network (AAR 2019a). These command centers are high-tech centers designed to offer decision makers access to real-time operational data from the network. These centers are also designed to visualize the data to aid decision-making. For example, manifest data provide detailed information about cargo contents and loads that control centers use to inform supply chain stakeholders after a disruptive event. This is critical, for instance, when an asset that is a nonredundant link enabling the movement of hazardous and sensitive materials undergoes a reduction in operational level.

4.4.4 R4: Recovery

Railroad infrastructure recovery is quantified in the context of operations. The primary goal of RR management after a disaster or disruptive event is to return the network to full service and operational levels in as little time as possible. Consequently, an objective measure of recovery includes two primary parameters: (1) the economic cost of improving robustness (or capacity) and the relative magnitude and size of the recovery, and (2) the amount of time required to return to the normal operating level. Owners and operators refer to RR network capacities (encompassing both throughput speed and tonnage) to quantify the level of recovery of an RR. Railroad managers ensure that their infrastructure can be measured at any point in time after a disruptive event by conducting detailed inspections in the field and updating internal rating systems that quantify the capacities based on those inspections (Moreu and LaFave 2012).

Railroads have made calculated efforts in recent years to accelerate the recovery process after disruptive events by increasing preparedness, streamlining damage assessment and repair efforts, and coordinating among RR stakeholders to better restart operations. Railroads have also extended the scope of recovery to include supporting relief efforts aimed at the recovery of affected communities. This includes, but is not limited to, supporting state and federal organizations by transporting clean water, food, shelter, and raw supplies to communities as well as removing debris. Railroad command centers work closely with weather personnel to monitor weather and anticipate extreme weather systems. When a forecasted disruptive event nears, RRs implement preventative measures to ensure that resources used in recovery efforts such as fuel, construction equipment, generators, and raw materials are protected and stockpiled, critical personnel such as engineers and inspectors are ready to take actions as soon as it is safe, and train and track equipment is protected (AAR 2019a). For example, RRs relocate locomotive and sensitive electronic track equipment to safe locations before extreme weather events (AAR 2019a). This helps to accelerate recovery efforts after a disruptive event because the RR owner/manager does not have to wait for damaged critical equipment to be replaced. After a disruptive event occurs, RR inspectors first assess the damage across the RR network and then collaborate with engineers and workers to initiate maintenance, repair, and replacement efforts. The efficiency of these two processes is integral to accelerating recovery. Consequently, when environmental or physical conditions make carrying out inspections and repairs hazardous during recovery, RRs leverage technologies to mitigate these hazards. For example, major railroads have used drones to conduct inspections of damaged RR infrastructure after natural disasters to keep workers safe (AAR 2019b). In addition, ground-penetrating radar is used to determine the amount of erosion and water-induced damage that exists under tracks after floods (AAR 2019b). Automated sensing technologies facilitate recovery efforts by providing owners/managers with the measured state of asset condition at its diminished operating level to rapidly inform the allocation of appropriate people and resources to the right places and determine when it has returned to full capacity.

Having established response protocols in place can also reduce the amount of time between a disruptive event and actionable response efforts. For example, the contents of cargo can drastically influence recovery efforts as operators work to return a RR to full capacity after a disruptive event. Railroads seek to prevent hazardous materials that are being transported from disrupting their operations. In the case of disruptions that involve toxic inhalation hazard (TIH) materials, protocol for full control of the scene is in place. At a network level, public relations between RR personnel and other stakeholders (e.g., FRA, DOT, states, local authorities, and communities) are essential to enable a quick recovery in the face of diminished operational levels. Efficient coordination among these stakeholders in response efforts allows for a faster recovery and aids in the formulation and implementation of effective contingency and emergency operations at the local level. Manifest data play a critical role in immediate recovery efforts because they allow workers to quickly assess the state of rolling stock and manage cargo throughout a supply chain network. Further, GPS and manifest data used together in a universal framework facilitate cooperation among RRs in the face of diminished operational levels. For example, the Federal Emergency Management Agency (FEMA), FRA, and DOT work with RRs to create and share emergency declarations, response plans, and regulatory waivers (AAR 2019b). In addition, RRs have detailed notification procedures that dictate who is provided information from an RR command center and when (CN 2019). Particularly in the case of TIH materials that may pose a serious public health concern, RRs will work directly with communities. For example, Organizations such as TransCAER (2019) facilitate RR-community relationships to prepare for and respond to TIH-based disasters.

4.5 CASE STUDY: RAILROAD RESILIENCE TO EXTREME FLOODING EVENTS

A case study illustrating the facets of how RRs respond to disasters (such as flooding) is presented with an emphasis placed on response in the context of the 4Rs. Flooding owing to hurricanes and heavy rainfall is one of the leading causes of weather-induced damage to RRs (AAR 2019b). In recent years, flooding and hurricanes in the United States have caused significant damage to RR operations, requiring considerable efforts to resume operations after disasters. Climate change is predicted to increase this occurrence and the severity of flooding events. In addition to damaging critical RR assets such as bridges, floods can wash away track ballast and damage electronic trackside equipment that controls system-level RR operations (e.g., signals). In the face of extreme water events that threaten both critical RR infrastructure and the communities they service, RRs have leveraged objective data sources and technologies to reduce the interruptions experienced during and after floods and hurricanes. This case study presents an overview of instances when the RR industry has done an exemplary job increasing network resilience in preparation and in response to extreme flooding events.

4.5.1 R1: Robustness

In anticipation of a likely increase in the intensity and frequency of storms, RRs are making concerted efforts to raise RR tracks and better detect flooding hazards to minimize the impact of floods on RR infrastructure (AAR 2019b). During the spring season of 2017, Missouri and surrounding areas in the Midwest experienced historic flooding, resulting in at least four locations setting new record crest levels (UP 2019a). Before the devastating Missouri floods, Union Pacific raised the level of 63 mi (101 km) of track and eight bridges along the Missouri River over a 6-week period (AAR 2019b). This directly increased the robustness of the RR network because sections of elevated tracks exhibited a lower reduction in operational level during and after the disruptive event. Across the RR industry, RR command centers also work closely with weather personnel to monitor weather and anticipate extreme weather systems. Railroads are augmenting these forecasting efforts by installing real-time flood detection technologies in regions that are prone to flooding. For example, high-water detectors are installed along tracks and are configured to automatically send warning notifications to approaching trains when flooding is a possible hazard (AAR 2019b).

4.5.2 R2: Redundancy and R3: Resourcefulness

Railroads carefully manage and prioritize their resources to ensure that resource flows enable proper decision-making strategies in real time in the face of flooding and storms. This is closely tied to efforts to increase redundancy in which RRs aim to ensure that operations continue in the event of disruptions by preparing for continued capacity across the network. Increasing resourcefulness and redundancy requires clear communication among stakeholders and an understanding of their immediate needs, a quantitative approach to prioritizing the repair of critical assets, and proper allocation of finite resources (including personnel, materials, and equipment) to be used in recovery.

Just 2 years after the devastating 2017 Missouri floods, the Midwest experienced even greater flooding because of the unprecedented melting of snow and ice that resulted in rivers rising to historic levels in over 40 locations across the Midwest (AAR 2019b). Critical RR assets (such as the Platte River Bridge) were submerged and tracks, including a critical section of the main transcontinental line in Nebraska spanning over 6,000 ft (1.83 km), were damaged (UP 2019a). In response to this disaster, Union Pacific managed the distribution of over 137,000 tons (124,284 metric tons) of materials (Table 4-2) across its Midwest network. During the recovery, Union Pacific identified Kansas City as having the greatest capacity to support rerouted trains (UP 2019a). However, the remaining available network (including hubs such as Kansas City) could not support the demand of all cargo that needed to be rerouted. Union Pacific communicated with stakeholders around the clock and coordinated with local organizations and government agencies to prioritize and deliver critical goods and services to consumers. For example, Union Pacific prioritized trains that ship grain used to feed animals and evacuate people out of the Missouri River valley to safe shelters (UP 2019a). Similar resourcefulness occurred in the aftermath of Hurricane Katrina in 2005 when Norfolk Southern was able to return its damaged RR network to operation in just 16 days by prioritizing resources (specifically nine cranes on barges) to repair the critical Lake Pontchartrain Bridge and lift 5 mi (8.05 km) of track from the water back onto this critical bridge (AAR 2019b).

4.5.3 R4: Recovery

When adverse physical or environmental conditions make carrying out inspections hazardous, RRs leverage technologies to mitigate these

Table 4-2. Resources Used by Union Pacific for Reconstruction Efforts after the 2019 Midwest Floods.

Technology/equipment	Quantity
Road bed material	>500 trucks carrying 65,000 tons (58,976 metric tons) of materials
Ballast and rock	Six large trains carrying 72,000 tons (65,317 metric tons) of materials
Excavators	21
Hyrail rotary dump trucks	20
Bulldozers	10
Hyrail excavators	9

Source: UP (2019a).

hazards and enable recovery to take place around the clock. As an illustrative example, during the 2019 Midwest floods previously described, elevated water levels, snow, and ice compromised the safety of inspection workers who needed to assess the damage across the RR network before engineers and workers could initiate maintenance, repair, and replacement efforts. To continue recovery in the face of this hazard, BNSF used drones to conduct inspections of damaged RR infrastructure to keep workers safe (AAR 2019b). In addition, ground-penetrating radar was used to determine the amount of erosion and water-induced damage under tracks after floods (AAR 2019b). Railroads also prepare critical resources before forecasted storms that might pose flood hazards, such as the 2019 Midwest floods. These preventative measures ensure that resources used in recovery efforts such as fuel, construction equipment, generators, and raw materials are protected and stockpiled, critical personnel such as engineers and inspectors are ready to act as soon as it is safe, and key train and track equipment is protected (AAR 2019a). As an example, RRs relocate locomotive and sensitive electronic track equipment to safe locations before extreme water events (AAR 2019a). This helps to accelerate recovery efforts after floods because RR owners/ managers do not have to wait for damaged critical equipment to be replaced.

4.6 CASE STUDY: AUTOMATED SENSING TECHNOLOGIES INSTALLED ON RAILROAD BRIDGES EXPOSED TO MULTIHAZARDS

The following subsections present case studies that illustrate the use of a number of diverse automated sensing technologies that can be permanently deployed on RR infrastructure exposed to multiple hazards. This section provides RR stakeholders with insight on how to map sensing data to quantitative assessment methods that are integral to resilience assessment, especially quantification of the 4Rs. Two case studies of the use of automated sensing technologies are presented. The first case study highlights rapid-to-deploy sensing technologies including wireless sensors and unmanned aerial vehicles (UAVs) to assess the behavior and capacity of timber rail bridges operating within their serviceability limits. The second describes the design of a permanent wireless monitoring system installed on the Harahan Bridge to assess the condition of critical bridge components exposed to multiple hazards, including seismic, vehicular collisions, and aging. The instrumentation strategy highlights the acquisition of data that allows the load demand and structural component capacity to be quantitatively assessed.

4.6.1 Rapid-to-Deploy Sensing Technologies to Assess Capacity and Serviceability of Timber Bridges

The use of monitoring systems to collect response data from RR bridges can provide objective data that can assist owners/managers in guiding resilience decisions and preventing unnecessary replacements after disruptive events (Moreu et al. 2017b). In other words, RRs can use objective information to show that an RR bridge performs safely under train loading and does not need postevent replacement. In addition, RR managers are typically aware of which assets in their networks are the most vulnerable. Consequently, objective data can be collected from network locations where there are signs of deterioration, decay, or low operational performance. Railroad departments that are responsible for bridge safety are interested in the local performance of specific bridge components (e.g., horizontal deflections in tall timber piles within timber trestles or vertical deflections in long spans). Monitoring local structural performance is of particular interest in the area of timber RR bridges because spans and piers repeat for hundreds of feet and their geometric and structural properties are, in general, similar between bridges. Understanding the performance of one component can shed insight into how similar components may behave in other parts of the bridge or in similar bridges.

Historically, two major response measurements are commonly made in bridges using automated monitoring technologies: strain and acceleration. Strain sensors (e.g., strain gauges) are used to measure the strain response of structural components. Accelerations provide insight into the global and local vibrations that are introduced from loads including train and ground motions. While both measures have proven effective in structural health assessments, RRs would much prefer to measure displacements. In particular, load–displacement relationships provide a basis for assessing the stiffness of a structural system, whether it is a local component or a global span. Displacement is also critical for checking that bridge responses remain within serviceability limits. Although displacement is potentially a more insightful structural parameter than strain and acceleration, displacement sensing is more difficult to implement in the field.

Displacement information collected from monitoring systems can be used by RR bridge owners/managers to help quantify robustness and make informed decisions about which bridges should be hardened. With respect to robustness, RRs can use displacements as an objective parameter to quantify the response of timber piers under train loads (which can be measured by wayside detection methods). This serves as an indicator of structural stiffness and system capacity. The American Railway Engineering and Maintenance-of-Way Association (AREMA) provides RRs with recommended practices for safe operations (AREMA 2019). AREMA recently included displacement measurements as an indicator of bridge safety. Specifically, AREMA recommends RR bridge owners/managers quantify excessive deflections and lateral/longitudinal movement that might require prompt closure of a bridge. However, the only explicitly cited limit is $L/250$ for net vertical deflection, where L is the span length. Past studies have shown that excessive displacements under train loads can be an indicator of unsafe operations and a trigger for maintenance and replacement operations (Moreu and LaFave 2011, Moreu and Spencer 2015, Uppal et al. 1990). Herein is a summary of a specific study in which displacement measurements are collected to quantify RR bridge robustness (Moreu et al. 2014, GAO 2018). Even though this section primarily discusses the structural response of a single timber RR bridge, the robustness of the asset has direct implications on redundancy, resourcefulness, and recovery within its greater subdivision and network.

Moreu et al. (2014) reports on a series of monitoring campaigns conducted between 2011 and 2013 that aimed to objectively quantify the reduction in capacity of timber RR bridges subject to train loads. The campaign specifically sought to measure maximum bridge displacements during train crossings. In total, 28 train crossing events were recorded and used to inform and relate displacement measurements to bridge conditions. Their study indicated that for similar timber RR bridges, transverse displacements were correlated to changes in structural condition observed in the field during construction. This reinforces the value of transverse displacements as an objective measure of capacity that can inform robustness and help owners/managers allocate resources for maintenance, repair, or replacement in recovery efforts.

Railroads are interested in using sensors to collect acceleration and displacement data from their bridges (Moreu et al. 2017b). The high cost of existing commercial sensors can hinder their adoption. Alternatively, low-cost sensors would enable RR owners/managers to more widely monitor critical assets and save money that can be redirected to repair and replacement efforts (in turn, increasing their opportunity to be more resourceful in their resilience strategies). The study by Moreu et al. (2014) highlighted that wireless sensing units can serve as a rapid-to-deploy sensing technology to collect measurements in the field after disruptions, especially in areas that are difficult to access with a lack of a fixed reference from which to measure. Figure 4-4(a) illustrates a wireless data acquisition system installed on timber bridges based on the Imote2 wireless sensing platform. The Imote2 was developed for field applications and is capable of power management and onboard processing, making it a suitable platform for use in structural health monitoring. This platform enables wireless telemetry and collects information about bridge displacements without the need for costly reference platforms. Linear variable differential transformers (LVDTs) were used to measure a timber support structure

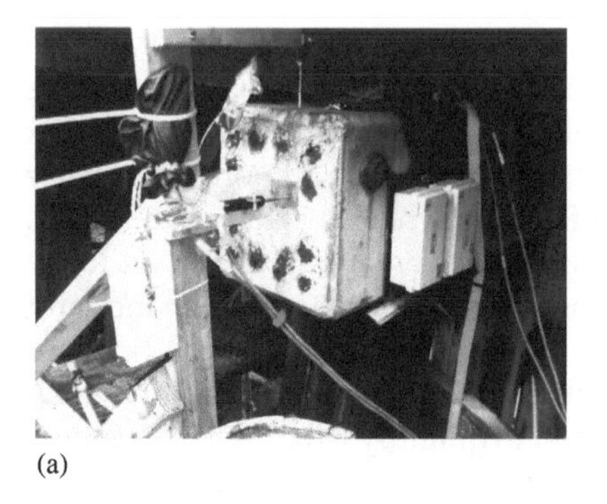

(a)

(b)

Figure 4-4. (a) Wireless sensing nodes to measure displacements (using LVDTs) and convert accelerations to displacements, (b) field validation for reference-free displacement measurements on a timber bridge structure.

that supports a deck plate girder span between a concrete pier and abutment [Figure 4-4(b)]. The LVDT provided a ground truth measure of vertical and transverse displacement at the top sections of the timber structure. The iMote2 also had a high-precision accelerometer on board that could be used to derive displacement through filtering and double integration methods (Moreu et al. 2014).

Based on the success of the iMote2 to acquire accurate measurement of bridge displacements, other researchers have explored even more effective displacement sensors for rapid deployment in bridges. Figure 4-5 shows a wireless sensor called *LEWIS* developed by Ozdagli et al. (2018). The *LEWIS* sensor comprises an inexpensive accelerometer to measure acceleration and interfaces with an Arduino-based embedded platform to convert the acceleration data into displacement data. This sensor is particularly attractive to RR managers because it is low cost compared with commercial sensors, operates wirelessly, collects high-resolution

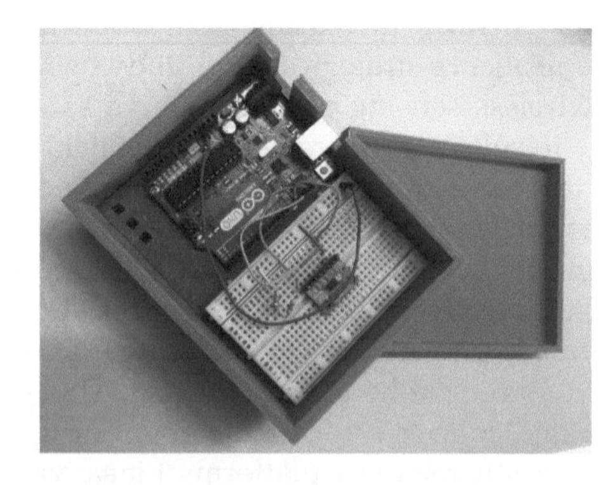

Figure 4-5. Low-cost LEWIS sensor for measurement of bridge displacements.

measurements, and is easy to use and install. This sensing node was used in the previously described monitoring campaigns of Moreu et al. (2014); the results show that the *LEWIS* sensors achieve equal or better displacement results than the commercial sensors in most of the test cases (Ozdagli et al. 2018, Gomez et al. 2019, Aguero et al. 2019, Ozdagli et al. 2019).

Although wireless sensors are easier to install than their wired counterparts, they still require labor to install. An ideal automated sensing technology would be noncontact and require no installation. Such a solution would be scalable in postdisaster assessment of bridge performance. Garg et al. (2019) adopted noncontact approaches to collect displacement data from bridges. In this work, a UAV system equipped with laser Doppler vibrometer (LDV) sensors was used to calculate the transverse dynamic displacement of a target. An algorithm was proposed to calculate the transverse dynamic displacement and compensate for measurement errors generated by the angular and linear movements of the LDV on the UAV. Figure 4-6 shows the setup of the field experiment conducted with an LDV mounted on a UAV. The results of this experiment were compared with measurements acquired by an LVDT contact sensor and indicate that the measurement errors in the LDV–UAV system were very small when compared with the LVDT measurements; the root mean square and peak errors between the calculated and the actual displacement values were less than 5% and 10%, respectively. The use of UAVs is increasing in the RR sector, especially to provide RR personnel critical information after an event, because of their scalability and the fact that they keep personnel away from unsafe field conditions.

(a)

(b)

Figure 4-6. (a) LDV equipped with UAV system for remotely measuring displacements in railroad bridges (LDV is mounted at the far end of the UAV shown in the lower right of the figure), (b) field test of a UAV measuring displacement of a moving target.

4.6.2 Permanent Wireless Monitoring of Large RR Bridges

This case study presents the design and implementation of a permanent wireless monitoring system installed on a long-span steel truss RR bridge to assess the condition of critical bridge components exposed to multiple hazards. Truss bridges used in rail networks (including the monitored Harahan Bridge) often comprise nonredundant fracture-critical components whose individual failure can result in bridge failure. Automated sensing technologies can be used to measure component structural responses that inform the safety and robustness of high-priority assets. Knowledge of the quantitative state of robustness can be used in efforts to prioritize the use of resources devoted to reducing the likelihood of failure of critical components through maintenance, repair, and replacement.

The Harahan Bridge (located near Memphis, Tennessee) (Figure 4-7) is a five-span cantilever truss bridge constructed in 1916 that carries traffic on two RR tracks and one pedestrian walkway over 4921 feet (1,500 m) across the Mississippi River. The Harahan Bridge is a critical piece of RR infrastructure that links Union Pacific's operations on the east side of the Mississippi River in Tennessee to operations on the west side of the river in Arkansas. That is, Union Pacific has just one crossing over the Mississippi River to access its operations in Tennessee and, hence, has a lack of internal network capacity to reroute rail traffic should the bridge fail because of external or aging hazards. The Harahan Bridge is exposed to diverse external hazards such as seismic activity (New Madrid Fault) and barge collisions with its piers, in addition to aging hazards such as weather-induced deterioration and train loading–induced fatigue accumulation. Given the critical role of the Harahan Bridge within Union Pacific's RR network, the bridge stands to benefit from automated wireless sensing

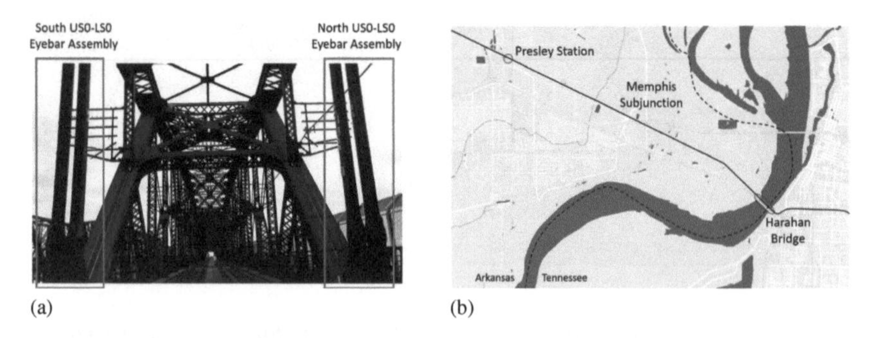

Figure 4-7. (a) North and south US0-LS0 eyebar assemblies on the Harahan Bridge, (b) location of the Harahan Bridge over the Mississippi River in Memphis (with Presley Station identified).

coupled with existing internal safety operations such as wayside detection, GPS, and manifest data.

In light of the aging and external hazards that the Harahan Bridge is exposed to, Flanigan et al. (2020a) undertook a long-term monitoring campaign of the Harahan Bridge to acquire data as inputs to a resilience analysis of the bridge. Global bridge responses are measured by accelerometers, whereas local responses are measured by strain gauges. A particular focus of the monitoring campaign was to assess the robustness of a fracture-critical bridge assembly: a set of eyebar truss elements designed to carry high tensile axial loads [see Figure 4-7(a)]. Given the high live load-to-dead load ratio of the bridge and the fact that trains use the bridge daily, fatigue of the eyebar truss assembly is a primary concern integral to both the element and bridge robustness. Along with bridge response data, additional data are collected from wayside detection stations near the Harahan Bridge. Trains operated by Union Pacific that cross the bridge have GPS units on board, whereas manifest data are recorded by the RR.

4.6.3 Wayside Detection, GPS, and Manifest Data

Union Pacific is increasing the resilience of its rail networks by widely integrating diverse data sources into the operation of its critical assets including the Harahan Bridge. These data sources, including wayside detection, GPS, and manifest data, are explicitly linked to the components of objective resilience (Table 4-1). Union Pacific is enforcing GPS-based PTC on its RR networks to prevent derailments owing to high train speeds, train-to-train collisions, and the movement of trains through misaligned track switches (UP 2018, 2019b). In addition, manifest data are collected for all trains that Union Pacific operates. Union Pacific operates the Presley Junction, which is the wayside detection station located closest to the Harahan Bridge. The Presley Junction includes a hotbox detection station that identifies faulty bearings to prevent seizure of bearings on the track. All trains that pass through the Presley Junction along the Memphis Subjunction pass through one (or more) of 17 WILD stations west of the Mississippi River. In addition to collecting information used to assess the robustness of rolling stock, the Presley Junction station collects data on a number of train features, including (but not limited to) the number of cars in the train, direction, speed, time, train ID, number of axles per car, number of locomotives, type of freight car and equipment type, axle loads, car weight, hunting oscillation index, car length, and distance between axles. WILD stations collect data for each train that passes through the Presley Station heading east toward the Harahan Bridge or west away from the bridge. Figure 4-7(b) identifies the location of the Harahan Bridge and the Presley Station relative to the Memphis Subjunction.

4.6.4 Automated Sensing Technologies

A permanent wireless monitoring system was instrumented on the Harahan Bridge in June 2016. The monitoring system architecture leverages the *Narada* wireless sensing node (Swartz et al. 2005) for time-synchronized data collection, processing, and transmission. Wireless monitoring systems emerged in the mid-1990s and have been growing in popularity as a lower cost and easily deployable alternative to traditional wired sensing (Kane et al. 2014). The wireless sensing nodes are controlled by a base station that initiates the data collection protocol, receives data from individual wireless sensing nodes, and transmits collected data to a server through a cellular link. A geophone is installed within the base station and is configured to continuously measure vertical velocity. When a set threshold is exceeded (indicating that a train is approaching), the base station sends a wake-up command to the wireless sensing nodes and initiates data collection. This data collection protocol ensures that the wireless sensing system remains in a low-power state when there are no trains crossing the bridge and activates the system each time there is a train event. The purpose of the automatic wireless sensing system is to measure the local and global behavior of key structural elements in the bridge. Figure 4-8(a) introduces the sensing instrumentation plan, which consists of eight strain gauges, six uniaxial accelerometers, and four triaxial accelerometers. Details about the deployed automated sensing technologies and sensors are provided in

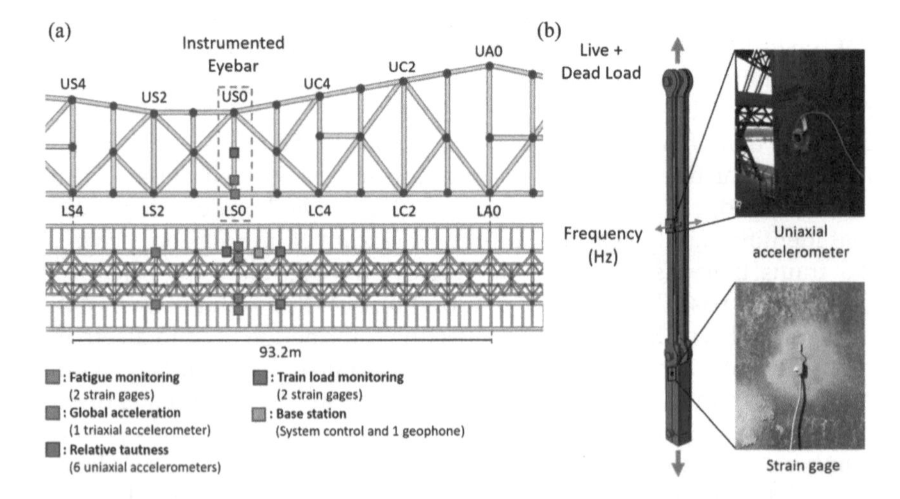

Figure 4-8. (a) Harahan Bridge instrumentation plan, (b) example of one strain gauge installed on the US0-LS0 box section and one uniaxial accelerometer installed on one of the north US0-LS0 eyebars.

Table 4-3. While the sensing system described herein is applied only to a small portion of the bridge, the instrumentation and data collection program can be naturally extended to any truss components.

Strain gauges are installed on the UC5-LC5 vertical hangers to gain insight into the locomotive and train car axle loads for train load monitoring and response behavior; these elements theoretically carry the entire train axle loads because they are only supported by upper and lower longitudinal chord elements. Because the UC5-LC5 vertical hangers are located adjacent to the monitored US0-LS0 members, time-synchronized strain measurements collected from both assemblies (together with train geometry information provided by wayside detection stations and manifest data) can be used to estimate the speed and direction of crossing trains. Triaxial accelerometers are installed along the north and south trusses to measure global acceleration in three directions. Flanigan et al. (2020a) describe the development of a finite-element (FE) model of the Harahan Bridge using CSiBridge (CSI 2016) that is intended to be used as a tool to determine the impact of failed bridge components on the overall structure (i.e., to identify fracture-critical components). The acceleration measurements, $\ddot{x}(t)$, collected by the triaxial accelerometers are transformed into the frequency domain (using the discrete Fourier transform), $|X(f)|$, and are used to calculate the Harahan Bridge's modal frequencies to verify the accuracy of the FE model's global response.

The main goal of the monitoring system is to use collected long-term measurements to assess the load demand and structural capacity of the fracture-critical US0-LS0 assembly. When used in a reliability analysis, long-term response data collected from the US0-LS0 assembly can quantify shifts in the statistical parameters characterizing the capacity and demand, and continuously monitor asset performance with respect to important failure limit states such as fatigue. The goal of reliability methods is to calculate the reliability index, β, corresponding to a failure limit state, which is a scalar measure of safety and serves as a quantitative metric for decision makers to explicitly assess robustness. Flanigan et al. (2020a) describe the full data-driven methodology for using collected strain and acceleration response measurements within a probabilistic fatigue analysis to assess the component (i.e., single eyebar element within the US0-LS0 assembly) and system (i.e., US0-LS0 eyebar assembly where all six bars have to fail for the system to fail) safety.

In the work presented by Flanigan et al. (2020a), uniaxial accelerometers installed on each of the six parallel eyebar plates in the out-of-plane (transverse) direction [Figure 4-8(b)] are used to calculate the relative tautness of each eyebar (i.e., the proportion of the total assembly load carried by each eyebar); the frequency of a vibrating eyebar is directly correlated to the total load it carries. Changes in the boundary conditions

Table 4-3. Data Sources Used for the Harahan Bridge Case Study Including Their Possible Use in Different Resilience Metrics.

Technology	Model	Quantity	Objective(s)	Resilience metric that might utilize the resulting information
Wireless sensing nodes	Narada wireless sensing node	10	Collect, process, and wirelessly transmit data	• Performance, Q • Time to recovery, T, $R3$ • Robustness, $R1$ • Resourcefulness, $R2$ • Redundancy, $R4$
Triaxial accelerometers	Silicon Design 2422-005	4	Measure asset response (vibrations) to loads; used for FE model updating via modal parameters	• Performance, Q • Time to recovery, T, $R3$ • Robustness, $R1$ • Resourcefulness, $R2$ • Redundancy, $R4$
Uniaxial accelerometers	Silicon Design 2012-002	6	Measure asset response (vibrations) to loads; assess relative tautness across fracture-critical parallel tensile eyebar elements	• Performance, Q • Time to recovery, T, $R3$ • Robustness, $R1$ • Resourcefulness, $R2$ • Redundancy, $R4$
Strain gauges	Hitec HBWF-35-125-6-10GP-TR with temperature compensation	8	Measure the strain response of structural components; assess train load; assess fatigue in metallic components	• Performance, Q • Time to recovery, T, $R3$ • Robustness, $R1$ • Resourcefulness, $R2$ • Redundancy, $R4$

Geophone	GeoSpace Geo-11D 4.5–380 VT	1	Trigger the wireless system to exit a low-power sleep mode and initiate data collection when a train is approaching	• Performance, Q • Time to recovery, T, $R3$ • Robustness, $R1$ • Resourcefulness, $R2$ • Redundancy, $R4$
Wheel impact load detection (WILD)	—	17	Measure load demand in the form of axle weights; identify out-of-round wheels	• Performance, Q • Time to recovery, T, $R3$ • Robustness, $R1$
Hot-bearing "hotbox" detectors	—	1	Identify faulty bearings to prevent seizure of bearings on the track	• Performance, Q • Time to recovery, T, $R3$ • Robustness, $R1$
GPS-based PTC	—	Network-wide	Geolocation (spatial mapping) and time information	• Performance, Q • Time to recovery, T, $R3$ • Robustness, $R1$ • Resourcefulness, $R2$ • Redundancy, $R4$
Manifest data and ACE	—	Network-wide	Collect detailed information about cargo, including customs bill ID, equipment ID, border crossing, shipping quantity and unit of measure, weight, and so on	• Performance, Q • Time to recovery, T, $R3$ • Robustness, $R1$ • Resourcefulness, $R2$ • Redundancy, $R4$

at the pin connections can cause the eyebars to carry different proportions of the load (Figure 4-9) and eyebars that carry a greater proportion of the load will fatigue at a higher rate. Accelerometers are selected over strain gauges because the proposed acceleration-based assessment calculates the *combined* dead and live load, whereas strain can be used to calculate only the live load. In addition, strain gauges are installed on the box section of the US0-LS0 assembly [Figure 4-8(b)] to measure the equivalent stress range assuming that the entire load across the assembly is applied to a single eyebar and calculate the number of stress cycles that occur during the monitoring period. The proportion of the total assembly load carried by each eyebar, P_i, $i = 1,2,\ldots,6$; the equivalent stress range, S_{re}; and the total number of cycles that occur during the monitoring period, N_t; are measured random variables that are used in Flanigan et al. (2020a) to calculate the load effect (i.e., demand) on the assembly. The result of the component-level reliability analysis carried out in Flanigan et al. (2020a) is shown in Figure 4-10(a) and reveals that the proportion of the total load carried by an eyebar influences its overall safety; Eyebar 2 carries the highest proportion of the total assembly load and consequently is the least robust over time. The evolution of the system-level reliability index is shown in Figure 4-10(b). In addition to illustrating that the reliability index serves as a scalar measure of safety and robustness that can be measured continuously, maintenance (i.e., eyebar realignment) occurring in the year 1950, 1980, 2000, or 2020 is simulated. Maintenance efforts lead to an instantaneous change in the trajectory of the reliability index and increase

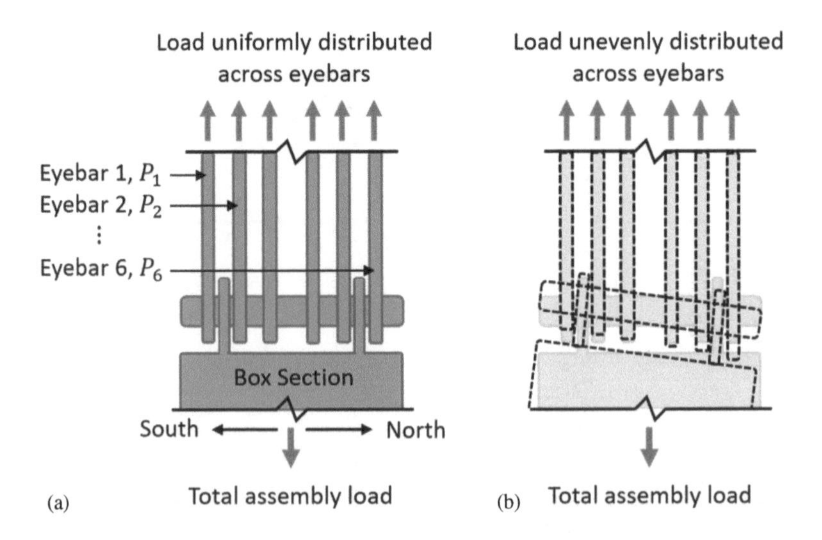

Figure 4-9. Illustration of (a) ideal boundary conditions, (b) change in boundary conditions due to an uneven distribution of loads across the eyebar assembly.

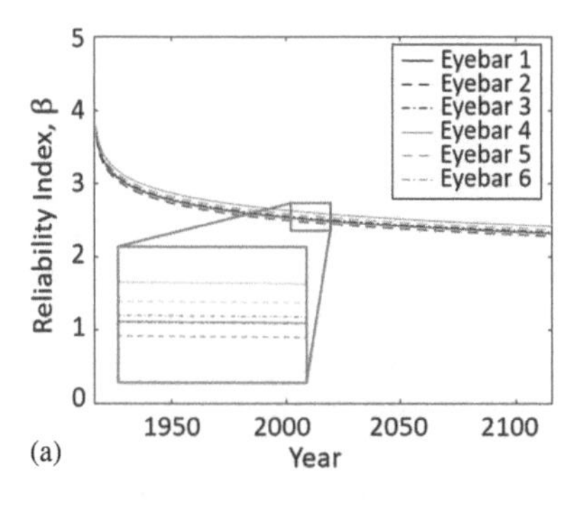

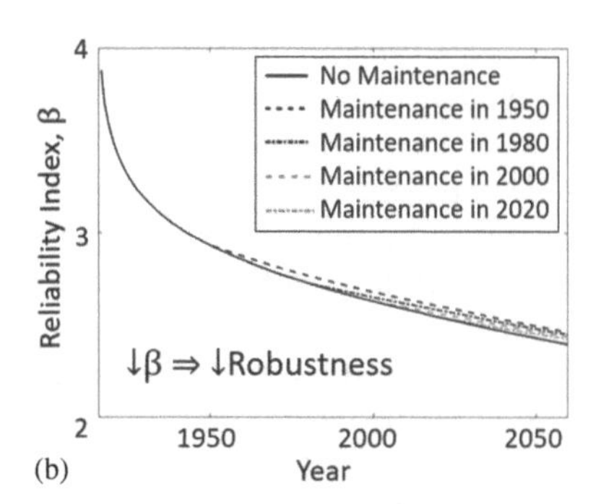

Figure 4-10. (a) Component reliability index evolution, (b) system reliability index evolution where maintenance to rebalance the eyebars occurs in 1950, 1980, 2000, or 2020. A 1% increase in annual train traffic is assumed for both analyses.

the future state of robustness. This result suggests that the use of reliability methods to quantify safety is a powerful objective approach that can aid decision makers in enhancing robustness.

4.7 PRINCIPLE OF RESILIENCE MONITORING

The main objective of this chapter is to explore how monitoring technologies can be used to obtain and update objective estimates of RR system resilience, including a mapping from data to resilience components. Lessons learned have relevance to all civil infrastructure systems, not just to RR systems. This chapter suggests that RMo should be performed as a subset of a complete RM effort to obtain an optimal result. The case studies illustrate that the results of different monitoring efforts can be used in diverse ways as inputs to compute the different components of system resilience and ultimately an estimate of the total resilience, Re, itself.

This chapter reveals that most of the information gathered through monitoring efforts interact together, both objectively and subjectively. While many of the variables that are necessary to estimate Re need to be estimated using objective methods such as numerical analysis of the system, it is sometimes more appropriate to estimate variables subjectively based on, for example, the experiences of the stakeholders.

The processes for executing comprehensive resilience monitoring of RR systems are currently not fully defined. Even though most of the ingredients necessary to quantify resilience through monitoring already exist, a principle of objectively performing integrated RMo in a comprehensive manner does not exist. Because of this, and based on the discussions provided in this chapter, the following principle of resilience monitoring (PRM) aims to provide such a comprehensive integration:

Comprehensive, integrated, and optimal RMo needs to contain, at a minimum, important elements of each of the two principal resilience components: performance, Q, and time to recovery, T.

Discussion: Note that performance, Q, and time to recovery, T, were shown to be the following: (1) Two necessary and sufficient components for completely defining any system resilience, including RR systems, according to all popular resilience definitions, including the NIAC definition (Gerasimidis and Ettouney 2022); (2) For a simple, yet demanding, formal resilience estimate, see Figure 4-1, which relates Re to Q and T.

It is clear that if the resilience monitoring system includes sufficient elements that can aid in an accurate estimate of Q and T, then it can produce the necessary and sufficient information to estimate Re and any of its subcomponents, such as the 4Rs. Figure 4-2 shows how RMo can be integrated within an overall RM process.

There are several important lemmas that need to be considered in conjunction with PRM. These lemmas are defined as follows:

Lemma 1: *Important elements of the PRM vary according to the scale and type of system under consideration.*

Discussion: Ettouney (2014 and 2022) argue that resilient systems scale from a single and limited asset (e.g., a building) to a large community (e.g., a city or a state). Hughes (2022) explored this resilience scale for rapid transit systems that include, but are not limited to, mass transit RR systems. While RMo should apply to systems of any size, system size does control the number, type, and nature of *important monitoring and data elements* that need to be included to ensure successful RMo implementation. The case studies presented in this chapter showcase this issue because the different elements comprising the monitoring systems were fairly different.

Lemma 2: *Traditional SHM is necessary, with certain conditions, but not always sufficient for comprehensive, integrated, and optimal RMo.*

Discussion: Traditional SHM (Ettouney and Alampalli 2011a, b, Lynch 2007, Lynch et al. 2016) has used many of the data types presented in Table 4-1. Consequently, SHM should naturally be included in any RMo effort. However, such inclusion comes with an important condition: its objective interrelationship with the identification of Q, T, and Re, or any of their subcomponents such as the 4Rs, needs to be well defined from the ground up. In addition, the degree of sufficiency of SHM results in how it identifies the Re components. If the information supplied by traditional SHM efforts is not sufficient, complementary information should be added using other numerical techniques or subjective estimates.

Lemma 3: *Visual inspection (VI) can easily and should be integrated in any effort for the RMo of the system under consideration.*

Discussion: VI has been the mainstay source of information in infrastructure management systems historically. This practice proved to be successful, accurate, and in many situations, optimal (Ettouney and Alampalli 2011a, b). Consequently, VI needs to be included, and perhaps expanded, in any RMo effort. Some development, and perhaps modifications of the practice, might be needed to expand the information gained from VI efforts so as to be of maximum utility in how it integrates with other data derived from RMo for use in defining parameters of the Re variable set.

4.8 SUMMARY AND RECOMMENDATIONS

North American RRs are considered to be among some of the best in the world, not only because of their economic impact and sustained health, but also because of their demonstrated resilience against hazards and disruptive events. Many RRs in North America prioritize safety and economy in operations to maintain and increase network growth. To safely maintain growth, volume, and operations across the country, RRs use objective data and information to inform their decisions before, during, and after disasters and major disruptive events. This chapter outlines the main considerations related to objective resilience in the context of North American RRs, where objective resilience comprises four key components: robustness, redundancy, resourcefulness, and recovery. The chapter details the explicit linkage between objective data sources and the 4 Rs to make them quantifiable properties of resilience in the context of RRs. These quantitative metrics for assessing resilience measure diverse parameters such as economic costs, efficiency of operations, freight throughput that needs to be maintained and recovered across a network, and safety of physical infrastructure components. An emphasis is placed on assessing operations and infrastructure from an engineering and operational perspective, which is highlighted by three case studies in which objective data sources are used to influence the shape of the operational level trajectory after a disruptive event occurs.

Despite the deliberate steps that are already being made (and in many cases enforced) across the RR sector to use state-of-the-art tools and technologies to enhance RR resilience from an objective perspective, decision-making remains a predominantly manual process that relies heavily on the experience of trained RR personnel. For example, even though this chapter describes the use of objective data sources such as automated sensing technologies, wayside detection methods, GPS, manifest data, and onboard monitoring to better quantify the 4Rs, these data are ultimately used to augment and better inform existing decision-making practices. More data and expansive analytical frameworks are direly needed to fully understand the interdependencies among the 4Rs and automate decision-making practices across the sector's diverse stakeholders. This will first require closing the gap between data and analytical approaches to assessing the 4Rs and their technical, social, economic, and political impacts.

4.8.1 Recommended Practices

This MOP aims to streamline resilience enhancement processes into a comprehensive and consistent RM paradigm. As argued throughout this MOP, the RM paradigm includes RMo as one of its components. This chapter offers many of the basic concepts and implementation details of RMo, especially as it pertains to RR systems. In closing, this section summarizes an annotated list of some of the most important RMo practices that can aid in producing resilient RR systems.

4.8.2 RR Assets

1. Better integration of collected data to quantify the state of RR assets in real time
   (a) Trains;
   (b) Soils and foundations (e.g., landslide, mud slide, rock fall events);
   (c) Track components (e.g., track buckling, rail break detection mechanisms, track fixation);
   (d) Ballast and sub-ballast;
   (e) Signals;
   (f) Support facilities;
   (g) Maintenance equipment; and
   (h) Other.
2. Integrate advanced warning mechanisms to detect hazards with built-in RMo systems.
3. Develop comprehensive RMo systems to accommodate predefined resilience acceptance thresholds; integrate these systems with pertinent decision-making systems.
4. Provide integrated capital improvement decisions with RMo systems.

4.8.3 RR Rolling Stock

1. Combination of onboard and trackside RMo systems for rolling stock deficiencies that may or may not be an indication of concerns from other RR assets
   (a) Wheels,
   (b) Axles, and
   (c) Bearings.
2. Any other improvement in rolling stocks that can have an impact and input informing RMo: (a) Increased Re of cars, and (b) Integrated rolling stock and RR assets.

3. Integrate RMo from a multimodal approach to better quantify objective resilience beyond the RR network.

4.8.4 RR Networks

1. Build network-based RMo systems that include asset-based RMo systems as subsystems;
2. Within the network-based RMo systems, include RMo systems for different resources as well as all recovery operations; and
3. There is a need for network-based objective resilience methods and processes that can take advantage of the results of RMo at a network level.

4.8.5 RR Operations and Management

1. Make full use of standard operating policies while planning RMo strategies;
2. Clearly delineate applicable hazards while planning RMo strategies;
3. Integrate different types of resource information with other types of information;
4. Streamline decision-making processes to optimally take advantage of RMo results according to Figure 4-2;
5. Identify and integrate human errors and human factors in the preventive and remedial solutions for a more comprehensive RMo strategy;
6. Improve, through reasonable exchange of RMo information and results, multidisciplinary coordination both vertically (within each organization) and horizontally (among different RR organizations and RR stakeholder groups);
7. Develop RMo so as to accommodate resilience communications from the ground up; this includes, but is not limited to, the following: (a) Messaging media (e.g., charts, videos, texting, mobile, and so on) that are compatible with the recipients (e.g., professionals, funders, legislature, owners, and so on), and (b) message content that is consistent with the objectives of the message (e.g., warning, decision support, scheduled information, and so on);
8. For more information about communications, see Ettouney and Alampalli (2017).

SUMMARY OF ACRONYMS AND VARIABLES

AAR = Association of American Railroads
ACE = Automated Commercial Environment
AREMA = American Railway Engineering and Maintenance-of-Way Association

BNSF = Burlington Northern Santa Fe
CFR = Code of Federal Regulations
CN = Canadian National
DHS = Department of Homeland Security
DOT = Department of Transportation
FE = Finite element
FEMA = Federal Emergency Management Agency
FHWA = Federal Highway Administration
FRA = Federal Railroad Administration
FTA = Federal Transit Authority
GAO = Government Accountability Office
GPR = Ground-penetrating radar
GPS = Global positioning systems
GRMS = Gauge-restraint measurement system
ID = Identification
LDV = Laser Doppler vibrometer
LEWIS = Low-cost, battery-powered, efficient wireless intelligent sensors
LiDAR = Light detection and ranging
LVDT = Linear variable differential transformers
MGT = Millions of gross tons
MPH = Miles per hour
NDI = Nondestructive inspection
NDT = Non-destructive testing
NIAC = National Infrastructure Advisory Council
NTSB = National Transportation Safety Board
PRM = Principle of resilience monitoring
PTC = Positive train control
Q = Performance
R1 = Robustness
R2 = Redundancy
R3 = Resourcefulness
R4 = Recovery
Re = Resilience
RM = Resilience management
RMo = Resilience monitoring
RR = Railroad
SHM = Structural health monitoring
T = Recovery time
t = Time
TIH = Toxic inhalation hazard
UAV = Unmanned aerial vehicle
UP = Union Pacific
VI = Visual inspection

VTI = Vehicle track interaction
WILD = Wheel impact load detection
α = Performance loss
β = Reliability index

ACKNOWLEDGMENTS

The authors would like to gratefully acknowledge Duane Otter for his valuable comments and suggestions. The Harahan Bridge case study presented herein was sponsored by DOT OST-R Award No. OASRTRS-14-H-MICH and Union Pacific Railroad. The opinions, views, findings, and conclusions are those of the authors and do not represent the official policy or position of the DOT/OST-R or Union Pacific. The timber bridge monitoring case study presented in this chapter was supported by the Center for Advance Research and Computing (CARC) of the University of New Mexico, Department of Civil, Construction and Environmental Engineering of the University of New Mexico, the Transportation Consortium of South-Central States (TRANSET), DOT under Project Nos. 17STUNM02 and 18STUNM03, New Mexico Consortium under Grant Award No. A19-0260-002, and TRB Safety IDEA Program under project "Measuring Behavior of Railroad Bridges under Revenue Traffic using Lasers and Unmanned Aerial Vehicles (UAVs) for Safer Operations: Implementation" (Project No. 163416-0399).

REFERENCES

AAR (Association of American Railroads). 2015. "Total annual spending." Accessed June 1, 2017. https://www.aar.org/Fact%20Sheets/Safety/AAR%20Annual%20Spending_2016%20Update_7.15.16.pdf.

AAR. 2019a. "How freight rail prepares for and responds to natural disasters." Accessed May 28, 2019. https://www.aar.org/article/freight-rail-natural-disasters/.

AAR. 2019b. "Resilient rail: How freight railroads prepare for and respond to extreme water." Accessed August 11, 2019. https://www.aar.org/article/resilient-rail-how-freight-railroads-prepare-for-respond-to-extreme-water/.

Aguero, M., A. Ozdagli, and F. Moreu. 2019. "Measuring reference-free total displacements of piles and columns using low-cost, battery-powered, efficient wireless intelligent sensors (LEWIS2)." *Sensors* 19 (7): 1549.

AREMA (American Railway Engineering and Maintenance-of-Way Association). 2019. "Structures, maintenance and construction."

In Vol. 2, Chap. 10 of *Manual for railway engineering*, 10-1-1–10-1-16. Lanham, MD: AREMA.

ASCE. 2006. *Hurricane Katrina: Performance of transportation systems*. Edited by R. DesRoches. Reston, VA: ASCE.

Barke, D., and W. K. Chiu. 2005. "Structural health monitoring in the railway industry: A review." *Struct. Health Monit.* 4 (1): 81–93.

Baylis, J., G. S. Gerstell, B. Scott, M. E. Grayson, C. Lau, and J. Nicholson. 2015. *NIAC Transportation sector resilience: Final report and recommendation*. Washington, DC: National Infrastructure Advisory Council.

Bemment, S. D., E. Ebinger, R. M. Goodall, C. P. Ward, and R. Dixon. 2016. "Rethinking rail track switches for fault tolerance and enhanced performance." *Proc. Inst. Mech. Eng., Part F: J. Rail Rapid Transit* 231 (9): 1048–1065.

Bruneau, M., S. Chang, R. Eguchi, T. O'Rourke, A. Reinhorn, M. Shinozuka et al. 2003. "A framework to qualitatively assess and enhance the seismic resilience of communities." *Earthquake Spectra* 19 (4): 733–752.

CFR (Code of Federal Regulations). 2018a. *Part 213—Track safety standards*. Title 49—Transportation subtitle B chapter 2—Federal railroad administration. Washington, DC: US Office of the Federal Register.

CFR. 2018b. *Part 237—Scheduling of bridge inspections*. Title 49—Transportation subtitle B chapter 2—Federal railroad administration. Washington, DC: US Office of the Federal Register.

CN (Canadian National). 2019. *Railroad emergency preparedness guide: Dangerous goods awareness level*. Montreal, QC: CN.

CSI (Computers and Structures). 2016. *CsiBridge 2016: Introduction to CSiBridge*. Berkeley, CA: CSI.

DOT (US Department of Transportation). 2015. "2012 economic census commodity flow survey." United States Census Bureau. Accessed June 1, 2017. https://www.census.gov/econ/cfs/.

DOT. 2018. "Statement on positive train control implementation." Accessed July 19, 2019. https://railroads.dot.gov/newsroom/statement-positive-train-control-implementation.

Ettouney, M. M. 2014. *Resilience management: How it is becoming essential to civil infrastructure recovery*. New York: McGraw Hill.

Ettouney, M. M. 2022. "Resilience management." Chapter 6 in *Objective resilience: Policies and strategies*, edited by M. M. Ettouney, MOP 146, 133–236. Reston, VA: ASCE.

Ettouney, M. M., and S. Alampalli. 2011a. *Infrastructure health in civil engineering: Theory and components*. Boca Raton, FL: CRC Press.

Ettouney, M. M., and S. Alampalli. 2011b. *Infrastructure health in civil engineering: Applications and management*. Boca Raton, FL: CRC Press.

Ettouney, M. M., and S. Alampalli. 2017. *Risk management in civil infrastructure*. Boca Raton, FL: CRC Press.

Falamarzi, A., S. Moridpour, and M. Nazem. 2019. "A review on existing sensors and devices for inspecting railway infrastructure." *J. Kejuruteraan* 31 (1): 1–10.

FHWA (Federal Highway Administration). 2006. "Freight facts and figures 2006." Accessed June 1, 2017. https://ops.fhwa.dot.gov/freight/freight_analysis/nat_freight_stats/docs/06factsfigures/index.htm.

Flanigan, K. A., J. P. Lynch, and M. Ettouney. 2020a. "Probabilistic fatigue assessment of monitored railroad bridge components using long-term response data in a reliability framework." *Struct. Health Monit.* 19 (6): 2122–2142.

Flanigan, K. A., J. P. Lynch, and M. Ettouney. 2020b. "Quantitatively linking long-term monitoring data to condition ratings through a reliability-based framework." *J. Struct. Health Monit. https://journals.sagepub.com/doi/10.1177/1475921720949965.*

FRA (Federal Railroad Administration). 2010. "Bridge safety standards." Accessed June 1, 2017. https://www.fra.dot.gov/eLib/details/L03212.

FRA. 2016. "Freight rail today." Accessed August 4, 2016. https://www.fra.dot.gov/Page/P0362.

GAO (US Government Accountability Office). 2009. "Freight rail security: Actions have been taken to enhance security, but the federal strategy can be strengthened and security efforts better monitored." GAO-09-243. Report to Congressional Requesters, United States Congress. Washington, DC: GAO.

GAO (US Government Accountability Office). 2018. *Grade-crossing safety: DOT should evaluate whether program provides states flexibility to address ongoing challenges.* GAO-19-80. Report to Congressional Committees. Washington, DC: GAO.

Garg, P., F. Moreu, A. Ozdagli, M. R. Taha, and D. Mascareñas. 2019. "Noncontact dynamic displacement measurement of structures using a moving laser Doppler vibrometer." *J. Bridge Eng.* 24 (9): 04019089.

Gerasimidis, S., and M. M. Ettouney. 2022. "On the definition of resilience." Chapter 1 in *Objective resilience: Policies and strategies*, edited by M. M. Ettouney, MOP 146, 1–24. Reston, VA: ASCE.

Gomez, J. A., A. I. Ozdagli, and F. Moreu. 2019. "Reference-free dynamic displacements of railroad bridges using low-cost sensors." *J. Intell. Mater. Syst. Struct.* 30 (9): 1291–1305.

Hughes, S. 2022. "Resilience of rapid transit systems: A practical outlook." Chapter 5 in *Objective resilience: Applications*, edited by M. M. Ettouney, MOP 149, 101–188. Reston, VA: ASCE.

Kane, M. B., C. Peckens, and J. P. Lynch. 2014. "Introduction to wireless structural monitoring systems: Design and selection." In *Sensor technologies for civil infrastructures: Performance assessment & health monitoring*, edited by M. Wang, J. P. Lynch, and H. Sohn, 446–479. London: Woodhead.

L.B. Foster Company. 2019. "Track monitoring solutions." Accessed September 23, 2019. https://www.lbfoster.com/en/market-segments/rail-technologies/solutions/rail-monitoring.

Li, C., S. Luo, C. Cole, and M. Spiryagin. 2017. "An overview: Modern techniques for railway vehicle on-board health monitoring systems." *Vehicle Syst. Dyn.* 55 (7): 1045–1070.

Liu, B., A. I. Ozdagli, and F. Moreu. 2018. "Direct reference-free measurement of displacements for railroad bridge management." *Struct. Control Health Monit.* 25 (10): e2241–e2256.

Liu, X., M. R. Saat, and C. P. L. Barkan. 2017. "Freight-train derailment rates for railroad safety and risk analysis." *Accid. Anal. Prev.* 98: 1–9.

Lu, P., R. Bridgelall, D. Tolliver, L. Chia, and B. Bhardwaj. 2019. *Intelligent transportation systems approach to railroad infrastructure performance evaluation: Track surface abnormality identification with smartphone-based app.* Rep. No. MPC 19-384. Washington, DC: University Transportation Center.

Lynch, J. P. 2007. "An overview of wireless structural health monitoring for civil structures." *Philos. Trans. R. Soc. London, Ser. A* 365 (1851): 345–372.

Lynch, J. P., C. R. Farrar, and J. Michaels. 2016. "Structural health monitoring: Technological advances to practical implementations [scanning the issue]." *Proc. IEEE* 104 (8): 1508–1512.

Lynch, J. P., M. M. Ettouney, S. Alampalli, A. Zimmerman, K. A. Flanigan, R. Hou et al. 2017. *Health assessment and risk mitigation of railroad networks exposed to natural hazards using commercial remote sensing and spatial information technologies.* Omaha, NE: USDOT and Union Pacific.

Magel, E., P. Mutton, A. Ekberg, and A. Kapoor. 2016. "Rolling contact fatigue, wear and broken rail derailments." *Wear* 366–367: 249–257.

McHenry, M., and J. LoPresti. 2016. "Field evaluation of sleeper and fastener designs for freight operations." In *Proc., 2016 World Congress of Railway Research.* Ferrovie dello Stato Italiane Group, Milan, Italy. Chicago: University of Illinois.

Moreu, F., H. Jo, J. Li, R. E. Kim, S. Cho, A. Kimmle et al. 2014. "Dynamic assessment of timber railroad bridges using displacements." *J. Bridge Eng.* 20 (10): 04014114.

Moreu, F., R. E. Kim, and B. F. Spencer Jr. 2017a. "Railroad bridge monitoring using wireless smart sensors." *Struct. Control Health Monit.* 24 (2): e1863–e1879.

Moreu, F., and J. M. LaFave. 2011. "Survey of current research topics—Railroad bridges and structural engineering." *Railway Track Struct.* 107 (9): 65–66, 68, 70.

Moreu, F., and J. M. LaFave. 2012. *Current research topics: Railroad bridges and structural engineering.* Newmark Structural Engineering Laboratory Report Series 032. Champaign, IL: University of Illinois at Urbana-Champaign.

Moreu, F., and B. F. Spencer Jr. 2015. *Framework for consequence-based management and safety of railroad bridge infrastructure using wireless smart sensors (WSS)*. Newmark Structural Engineering Laboratory Report Series 041. Champaign, IL: University of Illinois at Urbana-Champaign.

Moreu, F., B. F. Spencer Jr., D. A. Foutch, and S. Scola. 2017b. "Consequence-based management of railroad bridge networks." *Struct. Infrastruct. Eng.* 13 (2): 273–286.

National Coordination Office for Space-Based Positioning, Navigation, and Timing. 2018. "Rail." Accessed September 23, 2019. https://www.gps.gov/applications/rail/.

Ngamkhanong, C., S. Kaewunruen, and B. J. Afonso Costa. 2018. "State-of-the-art review of railway track resilience monitoring." *Infrastructures* 3 (1): 3–20.

NIAC (National Infrastructure Advisory Council). 2009. *Critical infrastructure resilience final report and recommendations*. Washington, DC: NIAC.

Nikolaidis, E., D. M. Ghiocel, and S. Singhal. 2004. *Engineering design reliability handbook*. Boca Raton, FL: CRC Press.

NTSB (National Transportation Safety Board). 1993. *Derailment of Amtrak train no. 2 on the CSXT Big Bayou Canot Bridge*. NTSB Rep. No. PB94-916301. Washington, DC: NTSB.

NTSB. 2013. *Metro-north railroad derailment*. NTSB Rep. No. NTSB/RAB-14/11. Washington, DC: NTSB.

Ozdagli, A. I., B. Liu, and F. Moreu. 2018. "Low-cost, efficient wireless intelligent sensors (LEWIS) measuring real-time reference-free dynamic displacements." *Mech. Syst. Signal Process.* 107: 343–356.

Ozdagli, A. I., B. Liu, and F. Moreu. 2019. "Real-time low-cost wireless reference-free displacement sensing of railroad bridges." In Vol. 8 of *Sensors and instrumentation, aircraft/aerospace and energy harvesting*, E. W. Sit, C. Walber, P. Walter, A. Wicks, and S. Seidlitz, eds., 103–109. Cham, Switzerland: Springer.

Rakoczy, A. M., D. E. Otter, J. J. Malone, and S. Farritor. 2016. "Railroad bridge condition evaluation using onboard systems." *J. Bridge Eng.* 21 (9): 04016044.

Riley, K. J. 2004. "Terrorism and rail security: Testimony presented to the Senate Commerce, Science, and Transportation Committee on March 23, 2004." Accessed September 15, 2021. https://www.rand.org/pubs/testimonies/CT224.html.

Strandh, V. 2017. "Exploring vulnerabilities in preparedness—Rail bound traffic and terrorist attacks." *J. Transp. Secur.* 10 (3–4): 45–62.

Swartz, R. A., D. Jung, J. P. Lynch, Y. Wang, D. Shi, and M. P. Flynn. 2005. "Design of a wireless sensor for scalable distributed in-network computation in a structural health monitoring system." In *Proc., 5th Int. Workshop on Structural Health Monitoring*, 12–14.

TransCAER. 2019. "Assisting communities and preparing responders for hazmat incidents." Accessed October 3, 2019. https://www.transcaer.com/.

UP (Union Pacific). 2017. "Safety technology and innovations." Accessed September 23, 2019. https://www.up.com/aboutup/community/safety/technology/index.htm.

UP. 2018. "Union pacific in Tennessee." Accessed September 23, 2019. https://www.up.com/cs/groups/public/@uprr/@corprel/documents/up_pdf_nativedocs/pdf_tennessee_usguide.pdf.

UP. 2019a. "How Union Pacific's actions helped customers and communities recover from the 2019 flood." Accessed September 1, 2019. https://www.up.com/aboutup/community/inside_track/2019-flood-recovery-6-28-19.htm.

UP. 2019b. "Positive train control." Accessed September 23, 2019. https://www.up.com/media/media_kit/ptc/about-ptc/.

Uppal, A. S., S. H. Rizkalla, and R. B. Pinkney. 1990. "Response of timber bridges under train loading." *Can. J. Civ. Eng.* 17 (6): 940–951.

USCBP (US Customs and Border Protection). 2019. "Automated commercial environment (ACE features)." Accessed September 23, 2019. https://www.cbp.gov/trade/ace/features.

Xu, X., Y. Lei, and F. Yang. 2018. "Railway subgrade defect automatic recognition method based on improved Faster R-CNN." *Sci. Program* 2018: 4832972.

CHAPTER 5

ROLES OF REMOTE SENSING TECHNOLOGIES FOR DISASTER RESILIENCE

ZhiQiang Chen, Margaret Glasscoe, Ron Eguchi, Charles Huyck, Bandana Kar

5.1 INTRODUCTION

Remote sensing (RS) acts as "eyes" looking at changes on Earth. As a data collection tool, it provides geospatial coverage of earth landscapes or objects, which include urban/rural communities and civil infrastructure assets. Because of its primary scale at a geospatial level (e.g., miles wide covering a megacity) and the modern RS sensor's high spatial resolution (e.g., at a submeter level per pixel), RS technologies have been widely considered in all phases of disaster management, including preparedness, response, and recovery.

In this chapter, the authors will first argue the relevance of remote sensing technologies to disaster resilience and suggest a general workflow for remote sensing–based damage detection (RT-DD) for civil infrastructure assets. To proceed, the background knowledge of remote (optical) sensing technologies, including their advantages and disadvantages, is described. Then, we review the emerging remote sensing technologies and their significant potential. With this basis, the RT-DD is formulated as a pattern classification problem, which elucidates the possible use of digital change detection, probabilistic classification, and crowd-based human identification for damage detection. Recent advances in big data and artificial intelligence will be remarked. Subsequently, the authors introduce three case studies to demonstrate the three methods. After concluding this chapter, the authors propose a possible framework for deploying remote sensing technologies in practice and suggest how to integrate RT-DD with practical disaster management.

It is noted that although the focus is on optical data and analysis methods, the general workflow and their essential relations to disaster resilience are universally applicable to different remote sensing methods. For processing nonoptical data [e.g., point-cloud processing or RAdio Detection and Ranging (Radar) data processing], specific methods are necessary. In terms of analysis methods, we focus on the modern use of statistical learning methods or machine learning methods, in which images are treated as the source data, then feature extraction methods, and pattern classification methods are applied in sequence. Emerging methods, such as the recent deep learning methods, are not focused but briefly introduced. In terms of damage detection, they essentially combine the feature extraction and classification steps into one large-scale neural-network-based model that is then trained on big data. In addition, this chapter includes a short introduction to deep learning methods for structuralizing emerging mobile images.

5.2 ROLE OF REMOTE SENSING TECHNOLOGIES FOR ENHANCING DISASTER RESILIENCE

5.2.1 Disaster Resilience

In the civil and infrastructure engineering community, the most widely adopted definition of resilience stemmed from the seminal work of Bruneau et al. (2003) in the context of earthquake hazard and engineering: "Seismic resilience is defined as the ability of social units (e.g., organizations, communities) to mitigate hazards, contain the effects of disasters when they occur, and carry out recovery activities in ways that minimize social disruption and mitigate the effects of future earthquakes." More impactful in this effort is the concept of "4Rs," which, as stated by Bruneau et al., are applicable to both social and physical systems including the following:

- *Robustness*: Strength or the ability of elements, systems, and other units of analysis to withstand a given level of stress or demand without suffering degradation or loss of function;
- *Redundancy*: Extent to which elements, systems, or other units of analysis exist that are substitutable (i.e., capable of satisfying functional requirements in the event of a disruption, degradation, or loss of functionality);
- *Resourcefulness*: Capacity to identify problems, establish priorities, and mobilize resources when conditions exist that threaten to disrupt some element, system, or other units of analysis; resourcefulness can be further conceptualized as consisting of the ability to apply material (i.e., monetary, physical, technological, and informational) and human resources to meet established priorities and achieve goals; and
- *Rapidity*: Capacity to meet priorities and achieve goals in a timely manner in order to contain losses and avoid future disruption.

It is arguable that the use of technologies can increase resilience performance at all different dimensions. In the original work of Bruneau et al. (2003), the authors proposed a system's diagram for quantifying infrastructure systems and community resilience; particularly in this diagram, it illustrates that "new approaches, such as the use of technological and decision support systems" can be incorporated to improve the resilience of an infrastructure system and a community. In the following, we especially elaborate on how RS technologies can enhance the resilience of (1) infrastructure asset as a monitoring and loss assessment technique, and (2) community resilience as a risk communication method.

5.2.2 Infrastructure Asset Resilience

When the physical infrastructure is targeted, the authors of this chapter state that in light of the existing research, developed technologies, and the emerging technology disruption, RS technologies have (1) proven strong relations with *Resourcefulness* and *Rapidness*, especially in community resilience at a geospatial scale. Because of their primary scale at a geospatial level and high spatial resolution, traditional RS technologies have been widely considered in all phases of disaster management, including disaster preparedness, response, and recovery. For example, one important application is loss estimation in the immediate aftermath of disasters through the use of visual interpretation or machine learning–based quantification. Particularly noteworthy is the latest advances in remote sensing platforms, for example, Google Earth Engine, which is enabling researchers and practitioners to develop big-data based analytics and smarter processing using the cloud computing infrastructure. With the ubiquitous use of mobile devices (e.g., smartphones), by taking advantage of modern communication networks (e.g., 4G and the upcoming 5G cellular networks), the disaster analytics is reachable to any person who is threatened by a disaster. This RS-based data and computing infrastructure are expected to be truly transparent to the end users and become a part of the socioeconomic fabrics of communities. Therefore, RS technologies as a whole have a strong relation to these two signature components of disaster resilience, resourcefulness, and rapidness. (2) Potentially beneficial influence on assisting with other technologies that provide *robustness* and *redundancy* to civil infrastructure assets. First, this impact is recognized in relation to two traditional technology arenas: structural health monitoring (SHM) and nondestructive evaluation (NDE) technologies. Both SHM and NDE methods have been long recognized as a viable approach to identifying structural states, inspecting structural damage, and providing diagnosis or prognosis for structural elements, devices, and subsystems in an engineering system's life cycle (Farrar and Worden 2006, Frangopol and Soliman 2016). These SHM/NDE techniques, by identifying, localizing, and quantifying damage states, directly contribute to implementing robustness and redundancy measures (by increasing the reliability of the system by replacing damaged elements or by adding additional elements or devices to the system) to the engineered system.

Given this, the authors recognize an emerging trend that RS is being exploited and integrated into the SHM/NDE fields. Two trends are recognized. Two potentially high-impact areas are recognized herein, which are both related to the advances of SHM and nondestructive evaluation (NDE) technologies. The first is to implement geospatial-scale SHM/NDE by integrating UAV-based remote sensing and ground-based sensing networks for the conjunct static damage detection and dynamic monitoring (Chen et al. 2018, 2019; Cheng et al. 2018; Yeum and Dyke 2015; Yoon et al. 2017). The second is the integration of dynamic building information modeling (BIM) and structural health information, leading to an enhanced digital-twin model for operation and management of engineered systems. In this aspect, the creation and updation of BIMs significantly rely on the emerging remote sensing technologies such as laser imaging, detection, and ranging (LiDAR) and 3D photogrammetric methods (Glaessgen and Stargel 2012, Qiuchen Lu et al. 2019, Shim et al. 2019).

With these relations as elaborated on RS technologies previously, we state that first, remote sensing provides an indispensable tool for *resilience monitoring* at global and regional scales. Second, remote sensing aids in resilient infrastructure asset operation and management by assisting and augmenting the existing and future SHM/NDE technologies.

5.2.3 Risk Communication and Community Resilience

With remote sensing–based data-derived products, such as hazard and loss mapping, they can be used to inform the public before a disastrous event, the immediate aftermath, or even in real time toward risk communication. The goal of risk communication is to provide adequate information about an impending hazard, its potential risks, and impacts as well as potential mitigation steps to aid communities in taking preparatory actions to reduce adverse impacts of the event (Gladwin et al. 2009, Krimsky 2007). Based on the definition of resilience, in addition to risk assessment, damage, and loss assessment, risk communication is considered a precursor to promote community resilience as it allows for a sharing of knowledge/information/lessons learned among stakeholders because they prepare to reduce impacts and prepare for future events and impacts (NRC 2012, UN/ISDR 2004). To this end, the *Sendai Framework for Disaster Risk Reduction 2015–2030* has identified the need to maintain a multihazard, multicultural, and people-centered forecasting and early warning system to provide risk information in a timely manner (UNDRR 2015).

A warning system required for risk communication comprises three components: (1) a detection subsystem that detects and/or forecasts the location and time of a hazard event, (2) an emergency management subsystem that focuses on determining the risk and impacts posed by the hazard that necessitates the dissemination of risk information, and (3) a public response subsystem that focuses on ensuring public understanding of messages and their subsequent response (Grabill and Simmons 1998, Kar et al. 2016, NRC 2012). In the United States, the Integrated Public Alert and Warning System (IPAWS) was established to disseminate standardized alert messages via multiple communication technologies (i.e., cell phones, radio, television) in multiple communication channels (i.e., text, pictures, sound) to impacted communities and beyond before, during, and after a hazard event promptly to reduce risk (FEMA 2012). Although technological advancements have made it possible to forecast hazard risks and impacts in near real time, public response to alerts and warnings is still influenced by (1) message content, and (2) the sociopsychological characteristics of the public (Kar 2018, Kar et al. 2016, Mileti and Peek 2000, NRC 2012).

While alert messages can be sounds received by sirens, the Wireless Emergency Alert messages received on personal cell phones tend to be 90 characters long (Kar et al. 2016). This limits the amount of risk information that the public expects to be part of a message, which includes information about the "nature of the disaster, impact zone, time frame and duration of the disaster, recommended actions, evacuation routes, when to take action, shelter location, how to obtain additional information, and a map of evacuation routes, shelters, and nearby hospitals" (Kar 2018). Even the rise of social media, specifically Twitter, as a risk communication platform does not provide information about the impacts of a hazard that essentially determines management and mitigation activities that can be taken up to move people out of harm or to improve response and recovery efforts (Liu et al. 2018). In light of the advances in remote sensing, geospatial, and computational technologies, it is now possible to determine risk, damage, and potential impacts as well as map them in near real time that can be used by both public and first responders as part of risk communication to increase awareness, preparedness capabilities, and reduce impacts, thereby enhancing resilience.

5.3 REVIEW OF REMOTE SENSING FOR DISASTER DAMAGE ASSESSMENT

5.3.1 Remote Sensing Basics

The inventory of remote sensing technologies today is much broader than ever before. In terms of sensing mechanisms, they include optical imaging (panchromatic, true-color, multispectral/hyperspectral, and

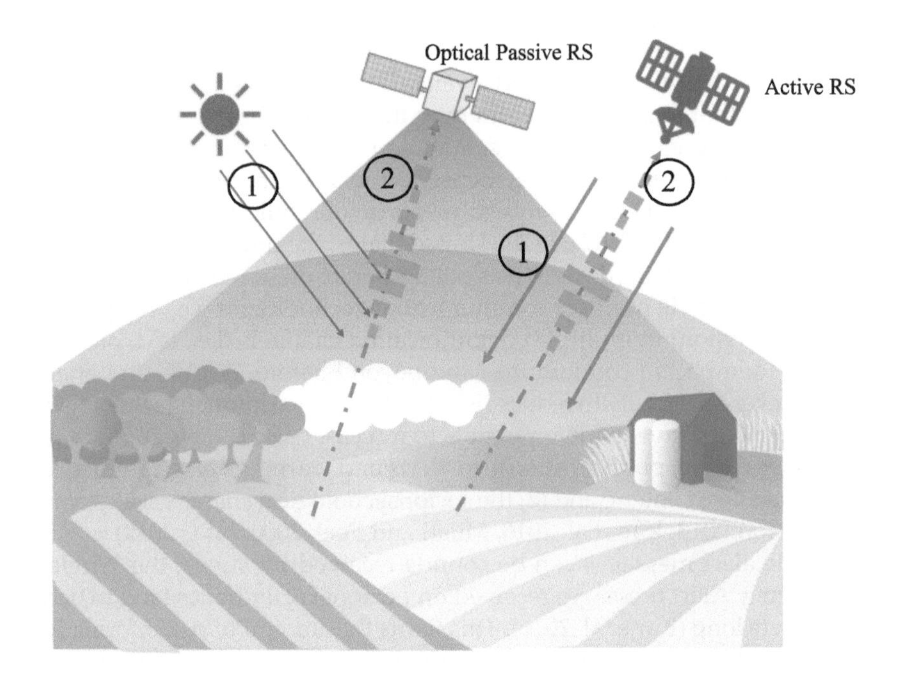

Figure 5-1. Traditional remote sensing platforms: optical passive RS and active RS.

thermal imaging), which passively relies on solar illumination. They also include active sensing methods, such as LiDAR (and radar), which actively emit electromagnetic waves and collect signals. Figure 5-1 illustrates the two mechanisms. In terms of platforms, they include spaceborne aircraft, airborne aircraft (manned), and unmanned aerial vehicles (UAVs).

It is noted that the knowledge domain of remote sensing is broad; readers may refer to a general textbook for the different remote-sensing principles (Rees 2013). The primary remote sensing data processing methods are instrumental prior to conducting any advanced interpretation (e.g., the damage detection problem in this book chapter), which are broad as well (Lillesand et al. 2015). With the ubiquity of mobile smart devices (e.g., smartphones and tablets) and micro- and small-scale UAVs (colloquially called drones), remote sensing to date can be even personally deployed, analyzed, and exploited in real time by individuals with a certain level of training.

Table 5-1 lists the primary representatives of these remote sensing technologies that are available to civic applications. To better understand

Table 5-1. Modern and Representative Remote Sensing Methods, Sensors, and Data Format.

Method	Representative sensors and owner	Imagery data format and notes
Very high-resolution optical sensors (submeter to meter/pixel)	Worldview-1, 2, and 3; GeoEye-1: Maxar SkySat: Planet Lab	$u(x, y, i)$: $u(\cdot)$ is measured as an earth-reflected radiance intensity value. Image format: GeoTiff or other image formats with embedded georeferencing for any pixel location (x, y) are commonly used. $i = 1$: scalar for panchromatic imagery, $i \in \{1, 2, 3\}$: RGB or true-color imagery $i \in \{1, 2, 3, \ldots\}$: RGB + one or more select near-infrared bands
Moderate- to high-resolution and multispectral optical imagery	Landsat series (currently active are Landsat 7 and 8): NASA/USGS Sentinel-2A, 2B: ESA SPOT series (5, 6, 7): Spot Image	u(x, y, b): $u(\cdot)$ is measured earth-reflected radiance intensity value. Image format: GeoTiff or other image formats with embedded georeferencing. Landsat 7: $b \in \{1, 2, \ldots, 8\}$ (with bandwidth from visible to short-wave infrared and thermal bands) Landsat 8: $b \in \{1, 2, \ldots, 11\}$ (visible to short-wave infrared and thermal) Sentinel 2: $b \in \{1, 2, \ldots, 12\}$ (visible to short-wave infrared)
Hyperspectral optical imagery	AVIRIS (airborne): NASA	AVIRIS: u(x, y, b): $b \in \{1, 2, \ldots, 224\}$ collected at 400 to 2,500 nm.

(*Continued*)

Table 5-1. (*Continued*)

Method	Representative sensors and owner	Imagery data format and notes
Small fixed-wing or multirotor UAVs	Commercial off-the-shelf (COTS)	$\{u_n(x, y, i) \mid n = 1, 2, \ldots\}$: image sequences with/without geotagging. Photogrammetric techniques (e.g., structural from motion, SfM) and software exist for obtaining 2D or 3D mapping products
Mobile smart devices	COTs (e.g., smartphones)	With accurate GSP (e.g., RTK station), high-precision mapping products can be achieved given densely overlapped and high-quality image sequences. Nonstructured mobile photos demand semantic processing and advanced analysis toward structuralizing.
Synthetic-aperture radar (SAR)	Sentinel-1A and 1B: ESA TerraSAR-X: DLR UAVSAR (airborne): NASA/JPL ICEye: ICEYE Oy	$u(x, y, s)$: $u(\cdot)$ is complex-valued (PolSAR) or real-valued (InSAR and so on) Signal processing is needed to process the Radar data, and most often, SAR polarimetry data are obtained PolSAR. In this case, s includes HH, HV, VV, VH four bands If bitemporal SAR data are used, by processing the phase data, interferometry SAR or InSAR is obtained. In this case, $s = 1$
Airborne LiDAR	High spectral resolution LiDAR: NASA	$p(x, y, z)$: point-cloud data; in general, (x, y, z) defines a world-coordinate at a collected point (hence, $p = 1$); if no collection, $p = 0$
Terrestrial LiDAR	COTS (e.g., Leica MultiStation)	

the nature of these data sets, the mathematical functional expressions of different imageries and their interpretations are summarized in Table 5-1. Among these remote sensing methods and their data products, a very large percentage of the remote sensing data and their products come from the traditional space- or airborne remote sensing platforms (e.g., satellites and air crafts) with optical sensors. Therefore, the majority of remote sensing data archived to date are optical data. To this date, there are several resources that provide freely available (with authorized accounts when necessary) remote sensing data, particularly during the aftermath of a disaster. Table 5-2 lists some centralized sources for acquiring remote sensing data that are provided by US-based government agencies. In the meantime, as personal or community-based remote sensing starts emerging, especially the use of GPS-enabled small UAVs, remote sensing data of disaster scenes are unprecedentedly ready to acquire; some of these postdisaster reconnaissance data sets are provided by researchers and recently are curated at the NSF's DesignSafe-CI cloud infrastructure.

Table 5-2. Remote Sensing Data Resources in the United States.

Name	Owner	Link	Main features
HDDS (Hazards Data distribution system)	USGS	https://hddsexplorer.usgs.gov/	Public-domain data and licensed (commercial); Optical and Radar imagery data; Event-based (2005 to current)
Earth science data systems	NASA	https://earthdata.nasa.gov/	Many sources, including satellites, airborne/field campaigns; Opensource APIs and tools
Google earth engine	Google	https://earthengine.google.com/	Historical and current Landsat and Sentinel-2 data; Cloud-based computing engine with image processing and machine learning algorithms
DesignSafe	NSF	https://www.designsafe-ci.org	Researchers collected remote sensing data Data types: Lidar point cloud, UAV imagery, ground-based (nonstructured) photos

5.3.2 Earth Observation Satellites for Disaster Response: From the Early 2000s

Several milestones of remote sensing technologies for earth observation (EO) can be recognized. These include the first orbital satellite launched in 1959 (US Explorer 6), the Landsat program starting from 1972 that ultimately leads to the largest program for Earth imagery acquisition from space in the human history (the latest launch was Landsat 8 in 2013), and the first real-time high-resolution Earth imagery acquired by the US KH-11 satellite. In 1977, the first electro-optical digital imaging was provided by the US KH-11 satellite system. Last, since 1997, the series of high-resolution EO satellites have been launched by the private sectors represented by DigitalGlobe (now Maxar), including IKONOS, Quickbrid, GeoEye-1, and the latest WorldView-1 to 4 series.

These modern satellite sensors are available to provide multitemporal imagery at very high resolution (VHR) at submeter spatial accuracy as well as other existing satellite systems that provide meter-level high-resolution (HR), such as the Ikonos and the SPOT-5. In addition to the high spatial resolution, some satellite sensors also provide wide bandwidths in the spectral domain and are operated to have shorter revisit periods (e.g., Worldview-3 and -4 have a revisit period of less than one day). Because of these capabilities, satellite images have been used as an unparalleled source of information for studying, monitoring, forecasting, and managing large-scale earth-surface phenomena, events, and human activities on our planet.

Although the adoption of remote sensing for disaster and emergency response can be dated back to the early days of satellite programs, it is as early as around the year 2000 that the research community in civil engineering started adopting statistical pattern classification or machine learning methods to exploit remote sensing data for postdisaster damage assessment. Many research papers are found in this period (Chiroiu 2005, Gusella et al. 2005, Kohiyama and Yamazaki 2005, Rathje and Crawford 2004, Rejaie and Shinozuka 2004). Partially, these research efforts were triggered by large-scale urban disasters and their immensely tragic consequence in terms of human and property losses around these years. These events include the 2004 Bam Earthquake in Iran, the 2004 Indonesian Ocean Earthquake and Great Tsunami, and the 2005 Hurricane Katrina in the United States. Second, during or after these disastrous events, remotely sensed satellite images were provided by different agencies and freely available to researchers around the world. Third, the early years of 2000 were also the golden age of statistical machine learning and machine vision methods. Besides technical papers, technical textbooks in these arenas were published (Bishop 2006, Chan and Shen 2005, Shawe-Taylor and Cristianini 2004), indicating the maturity of these methods and readiness for their adoption in the civil engineering communities. During these times, researchers developed advanced Bayesian learning methods, graphical modeling theories, statistical kernel methods, and advanced image-based feature extraction methods, which were all ready for application.

With the aforementioned orbital sensing systems, individual urban infrastructure (e.g., buildings, roads, bridges) and their corresponding details that characterize them (e.g., boundaries or roof details) are visually identifiable and reasonably detectable in the obtained RS images. Therefore, the researchers mentioned previously in civil engineering attempted to adopt these machine learning and vision methods and expected to develop an autonomous (or safe to say, semiautonomous) method for rapid RS-based damage and loss mapping. These endeavors were expected to assist in performing immediate rescue and reconnaissance in disaster-stricken areas as well as to provide useful information for improving hazard reduction methods for designing or maintaining engineering structures. In the following, a previous event is used to illustrate such potential for using RS data for postdisaster damage assessment.

The Great Sumatra Earthquake and the subsequent Great Indian Ocean Tsunami caused by the uplift of the ocean floor, occurred on December 26, 2004, claimed more than 300,000 lives, and caused billions of dollars of property losses across 12 countries bordering the Indian Ocean. This was clearly a large-scale disaster. Within only the area of Aceh Province of Indonesia, about 1.3 million urban buildings were damaged because of the devastating effects of the tsunami. Different from other natural hazards, most affected areas were hit by the earthquake ($M_w = 9.2$) and subsequently stricken by the tsunami.

In Banda Aceh and other affected areas in Indonesia, the majority of built structures can be grouped into three types: (1) residential buildings constructed with wood framing and clay tiles or thin corrugated steel sheets for roofing, (2) nonengineered concrete-framed structures with unreinforced masonry infill, and (3) engineered reinforced concrete buildings. According to several reconnaissance reports (e.g., Ghobaraha et al. 2006, JSCE 2005), the damage in the city of Banda Aceh was the result of both the seismic shaking and the subsequent tsunami. Different from seismic shaking and its dynamic effects on structures, tsunami waves apply hydrodynamic pressures, buoyancy, uplift, or scour to structures. Hence, the damage patterns manifest very differently from typical seismic-alone damage.

In Figure 5-2, three rows of urban buildings in the Banda Aceh area from different sources are provided. In the first row (a1–a3), a typical building collapse was observed, which was due to the poor resistance of the buildings to seismic shaking. In the next row of pictures (b1–b3), buildings were primarily destroyed by the tsunami waves. In Figure 5-2(b3), only the

Figure 5-2. Urban buildings struck by the Sumatra Earthquake and the Great Tsunami: (a1–a3) Three reinforced concrete buildings in Banda Aceh collapsed during the earthquake; (b1–b3) show three buildings with complete or near-complete collapse owing to tsunami waves and flooding; (c1–c3) damaged buildings as a result of the tsunami impact with different effects: (c1) ground-level columns damaged by flooding pressure and piled debris, (c2) a survived building with the ground floor soaked in water, and (c3) a building with loss of support of its shallow foundation owing to water erosion.
Source: All illustrations created based on sample images in (a1–a3) JSCE (2005), (b1–b3) USGS (2008), and (c1–c3) Ghobaraha et al. (2006).

ground story was left, with the second story almost completely gone. It is noticeable that large amounts of debris were found piled around the buildings, which were left by the receding water and may not be solely from the collapsed buildings in situ. In Figure 5-2(c1–c3), readers may observe three particular damage patterns caused by the tsunami, which represent three special situations that satellite sensors on the orbit cannot capture. In (c1), the ground-level columns were apparently damaged by flooding pressure and the accumulated debris pressure; in (c2), a building stood, however, was soaked in water extensively; and in (c3), the shallow foundation of the building was eroded by water. All these effects could not be captured by satellites with nadir- or off-nadir view. It will be commented later that for situations in Figure 5-2(c1–c3), the latest UAV-based remote sensing operated at a low altitude can offer a unique approach to remote sensing of these damaged structures aided with modern high-resolution cameras and GPS.

It is noted that on December 28, 2 days after the event, the Quickbird imagery was available. These high-quality images clearly displayed the extent of the flooding, the severity of urban infrastructure damage, and the widespread devastation of the city and its surrounding areas. The first author of this chapter licensed several pairs of bitemporal Quickbird images from the Pacific Disaster Center (PDC 2008). These images covered the Banda Aceh and its surrounding areas and were pan-sharpened with a spatial resolution of 60 cm/pixel. These images are very large; two example bitemporal image patches are re-created with a downgraded resolution and are shown in Figure 5-3. Note that their original approximate extents are indicated on a small browser image.

In the first pair of bitemporal patches (P1), the neighborhood shown was close to the shore and was entirely destroyed by the tsunami waves with randomly cluttered debris illustrated in the postevent image. In P2, visual identification reveals that most of the buildings in this region stand intact after the disasters. However, severely inundated land was observed. Therefore, potential severe damage to buildings may not be observed from the satellite images.

If the geometric and spectral intrinsic of the satellite sensor and other environmental factors are held constant, the topographical appearance of urban buildings in satellite images manifests as individually distinct in terms of (1) geometric shapes, and (2) reflectance properties. The 3D geometric shape of an urban building determines its 2D projection in the satellite image. The reflectance properties of buildings vary spatially differently as different construction materials were used, which cause different spectral compositions under the solar illumination when captured by optical sensors.

Because of the nature of optical imaging, many physical and environmental factors not related to the buildings have a considerable effect on the topographical appearance of the buildings. These factors, to name a few, include sensor tilts, solar illumination variations, atmospheric disturbance, and weather changes. For example, sensor tilts can project the building geometrically different in the image; solar angles can produce different shadow patterns to the same building; atmospheric disturbance may lead to diffusive artifacts in the image.

One may note that VHR satellite images can display the following damage patterns: (1) Partial or complete structural collapse, wherein a part of or a whole of the building reduces to debris, and (2) distortion of the geometric configuration of buildings, which may include the disappearance of the original boundaries or the appearance of new segments or boundaries.

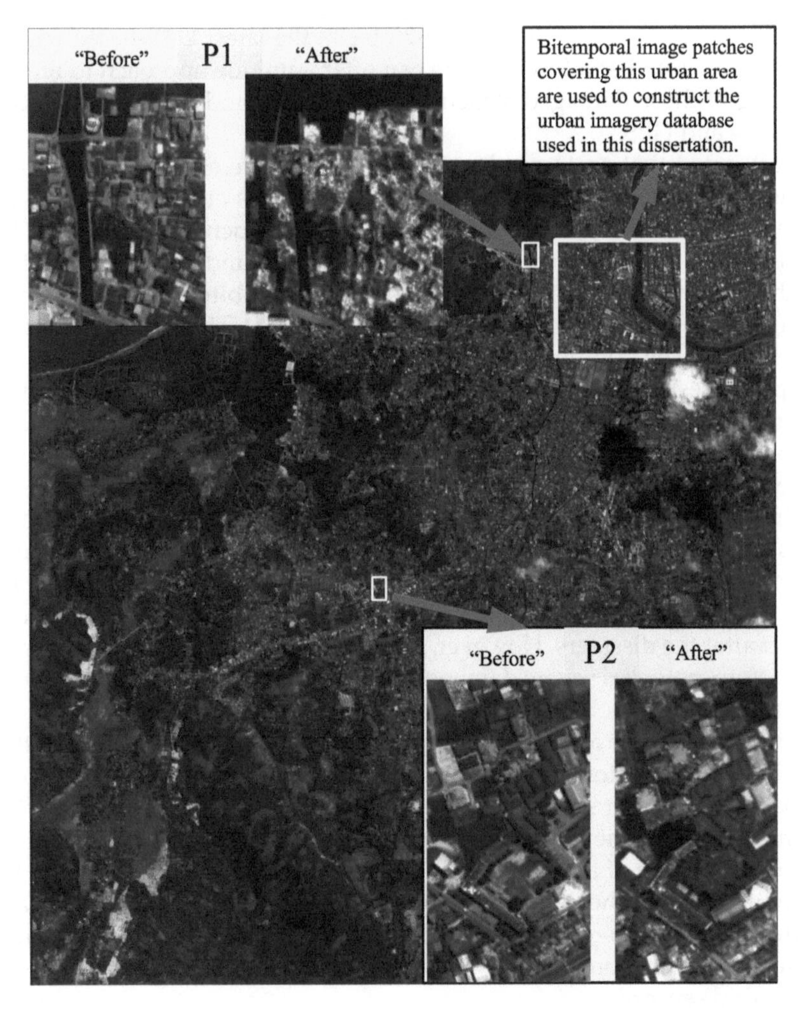

Figure 5-3. Two pairs of bitemporal image patches (P1 and P2) cropped from a pair of large-scale satellite images captured before and after the 2004 Indonesia Ocean Tsunami. The pre-event image was captured on June 23, 2004, and the postevent one was captured on December 28, 2004.
Source: Created based on resolution-downgraded images licensed through the PDC (2008).

It is also recognized that compared with photographs taken at the ground level as well as other on-site data such as the collection of eyewitness information and detailed manual mapping of damage patterns, many types of structural damage characteristics cannot be identified from satellite images. However, the benefits of satellite imagery are unparalleled when considering the following two facts: The ability to capture disaster scenes on a large scale. In the case of an earthquake that primarily impacted a small city or region (e.g., the Bam, Iran, earthquake in 2003), most of the loss and damage were spread within a (4×4) km^2 area; in the South Napa Valley Earthquake in 2014, most building damage was confined within the downtown Napa in about a 5×6 km^2 area. Such an extent can be largely captured in a single $(7{,}000 \times 7{,}000)$-pixel or $(8{,}000 \times 10{,}000)$-pixel image with a spatial resolution of 60 cm/pixel. For disasters like the 2004 Indian Ocean Earthquake and Tsunami across multiple nations or the 2011 Tōhoku Earthquake and Tsunami in Japan that flooded more 500 km^2 of coastal lands, a number of satellite images are needed to provide broad coverage for the extent of damage. Modern satellite imaging systems well accommodate this requirement. For example, the Worldview-3sensor, when operated with an off-nadir angle in the range of 30 degrees, can cover a geospatial area up to 66.5 km $\times$ 112 km in a single pass (SIC 2021), (2) Timeliness of satellite imaging. Modern commercial satellite systems usually feature a short revisit period. The result is that within one or a few days following the disaster, satellite imagery can be available. Although the processing and computing of such large satellite images are very demanding, this should not be viewed as obstacles, as computing technology continues advancing, and high-performance cloud infrastructure for satellite processing becomes normal to date. Compared with ground-level reconnaissance reports, which often require months to generate, the time consumed by image processing and computing is negligible.

5.3.3 Emerging Remote Sensing Technologies

The previous example shown in Figure 5-3 implies that structural damage in built objects is often not sensible by orbital satellites with optical sensors. Disaster scenes are far more complex, however. In Figure 5-4(a), one scene from the 2010 Haiti Earthquake is shown, in which the damage from the severely damaged community is spatially dynamic in terms of the terrain complexity that challenges any orbital or airborne sensors. Figure 5-4(b) illustrates another complicated imaging situation, wherein damage to bridges is mostly found in the substructure of bridges owing to either seismic or tsunami wave impact (JBEC 2011). In these two situations, any conventional spaceborne remote imaging approach is not feasible to capture the damage scene rapidly. The recent emergence of small UAVs that fly at a low altitude provides a flexible and cost-effective approach to providing multiview imagery of objects, potentially enabling 3D scene reconstruction (Chen 2014).

Several researchers have reported this UAV-based remote sensing for several disaster scenes such as hurricanes and earthquakes as a rapid and alternate remote sensing approach (Murphy et al. 2008, Pratt et al. 2009). One recent UAV-based disaster-scene remote sensing work was conducted

Figure 5-4. (a) Several damage scenes from the 2010 Haiti earthquake with complex terrain, and (b) a damage scene showing debris impact on a bridge pier in the aftermath of the 2011 Japan Tsunami.
Source: (a) Based on Internet pictures, (b) modified based on JBEC (2011).

by the first author in the aftermath of an EF4 tornado that occurred in eastern Kansas on May 28, 2019, causing considerable damage in Linwood, Kansas, and surrounding areas. The tornado ultimately resulted in 18 injuries and no fatalities. Using a DJI Phantom 4 UAV, high-resolution images were collected through the use of a multispectral camera and a 4K RGB camera.

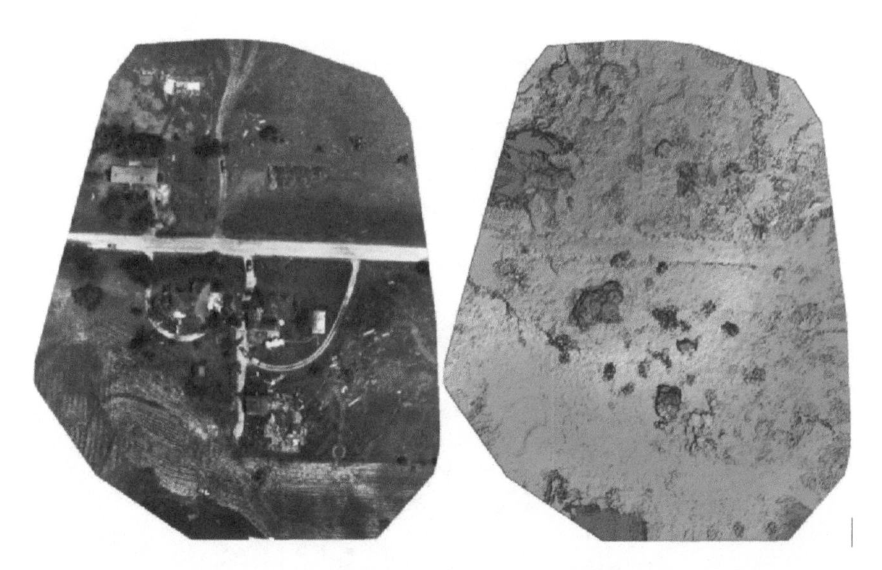

Figure 5-5. Pix4D processing products: orthomosaic and digital surface model (DSM) using the UAV images near 1663 N 1300 Road, Lawrence, Kansas. It is noted that the DSM model can be used to assess the volume change of a structural or vegetation object if necessary for quantitative tornado damage assessment.

With the images obtained from the UAV, preprocessing using Pix4D (Pix4D 2021) mapping was conducted. Figure 5-5 shows the intermediate mapping products and the final reconstructed 3D mapping. It is noted that to achieve a rapid product, the low-resolution 3D mapping option was selected. Even so, the achieved resolution of about 1.4 cm/pixel for both locations was ensured.

The low-resolution products were further exported into Google Earth. Figure 5-6(a and b) show the "before" and "after" using existing satellite images and the derived mapping product for the location of 20361 Linwood Road, Linwood, Kansas, 66052. Figure 5-7(a) shows a detailed "zoom-in" illustration from using the 3D mapping product near the address number of 1663 N 1300 Road, Lawrence, Kansas, 66046. It is of interest to note that the falling trees damaged both residential buildings. For the building on the right, a ground photo was taken as well [Figure 5-7(b)], which manifested the severity of the damage and confirmed the aerial observation. More than what is shown in the ground photo, the aerial mapping shows the relative position of the damaged area to the roof system.

5.3.4 Relation to Other Hazard and Loss Estimation Methods

Related to the category of hazard and loss assessment, other geospatial tools (software or system platforms) exist in practice and in the literature.

(a)

(b)

Figure 5-6. Pre- and post-tornado disaster scene comparison using the existing satellite image and the produced mapping product.
Source: Courtesy of Google Earth Pro.

(a) (b)

Figure 5-7. Aerial and ground comparison of a residential building at Lawrence, Kansas: (a) the zoom-in illustration from the UAV mapping; and the building in the right was captured in (b) by a smartphone at the ground level.

Some of them are based on simplified and predictive modeling methodologies, some combine probabilistic modeling and integrate with real-time networked sensing data, and some use advanced physics-based computational modeling. A few examples are listed as follows:

1. Hazus-MH: Hazus is a nationally applicable standardized methodology and a GIS-based software package that contains models for estimating potential losses from earthquakes, floods, hurricanes, and tsunamis. It uses empirical vulnerability functions and exposure data to estimate structural losses at a census block scale and provides important loss estimates before a disaster (Schneider and Schauer 2006).
2. USGS seismic shaking maps: ShakeMap is a product of the USGS Earthquake Hazards Program in conjunction with the regional seismic networks. ShakeMaps provide near-real-time maps of ground motion and shaking intensity following significant earthquakes (Frankel et al. 2000).
3. NOAA hurricane simulation tools: NOAA provides a set of toolkits that use advanced atmospheric and climate models for hurricane behavior simulation, which provides geospatial prediction of hurricane movement and intensity (Walsh et al. 2015).

In general, these geospatial tools serve as either predisaster preparedness planning or during-/postdisaster awareness of hazard intensity (as a proxy to the potential loss) and estimates of potential losses. They are different from remote sensing–based damage mapping in that the latter provides data reflecting real and consequential damage owing to the disaster. In practice, if no remote-sensing data capture disaster scenes, the forecast or predictive hazard/loss-estimation products provide substantial assistance in decision making.

5.4 RS-BASED DAMAGE DETECTION AND COMPUTING METHODS

5.4.1 Problem Definition

Given the potential of satellite imaging for urban disaster response and management, it is important to understand how urban structural damage can be described in satellite images. One viable strategy is the use of categorical damage grades, such as those shown in Figure 5-8 (EMS, 1998). The illustration shows five damage grades, which include descriptive language regarding the structural or nonstructural components for each grade. This damage grading system is often used as a reference during postdisaster structural damage reconnaissance.

Classification of Damage to Masonry Buildings	
	Grade 1: negligible to slight damage (no structural damage, slight non-structural damage) Hairline cracks in very few walls; fall of small pieces of plaster only; fall of loose stones from upper parts of buildings in very few cases
	Grade 2: moderate damage (slight structural damage, moderate non-structural damage) Cracks in many walls; fall of fairly large pieces of plaster; partially collapse of chimneys.
	Grade 3: substantial to heavy damage (moderate structural damage; heavy nonstructural damage) Large and extensive cracks in mos walls; roof tiles detach; chimneys fracture at the roof line; failure of individual non-structural elements (partitions, gable walls)
	Grade 4: very heavy damage (heavy structural damage, very heavy non-structural damage) Serious failure of walls; partial structural failure of roofs and floors.
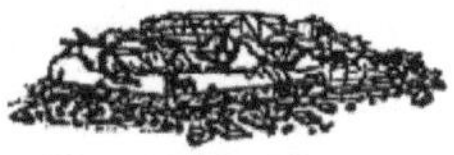	Grade 5: destruction (very heavy structural damage) Total or near total collapse

Figure 5-8. Damage grading and description.
Source: Modified based on EMS (1998).

There is limited work in the literature addressing the identifiable levels of damage in satellite images. In Yamazaki et al. (2005), the authors manually conducted experiments by applying visual inspection to Quickbird images, in which more than ten thousand buildings in Bam, Iran, were labeled with different damage levels (Yamazaki et al. 2005). In their experiment, a reduced 4-level damage grading system was used: Damage Level 1—buildings with no or slight damage, which are equivalent to Grades 1 and 2 in Figure 5-7; Damage Level 2—buildings surrounded by debris (Grade 3 in Figure 1-10); Damage Level 3—buildings that partially collapse (Grade 4 in Figure 1-10); and Damage Level 4—buildings that completely collapse (Grade 5 in Figure 1-10). After comparing the visual identification results with a ground-level reconnaissance report, they concluded that consistency was achieved over most of the buildings in the images. However, inconsistent cases were found, especially when deciding the intermediate damage levels (e.g., Damage Levels 2 and 3).

It is not justifiable nor discernible if a damage grading system with more than three levels can be used for the case of remote sensing data. The primary reason is that traditional remote sensing data is not effective in capturing the 3D appearance of ground buildings. To this end and for many remote sensing–based damage assessment efforts, a more reduced damage grading system is used, which includes three damage levels:

- *Minor damage*, which corresponds to Grades 1 and 2 in Figure 5-7, that is, structures are intact or have only a slight loss of structural features (e.g., edges like boundaries and homogeneous roof regions);
- *Moderate damage*, which corresponds to Grades 3 and 4 in Figure 5-7, that is, the partial collapse of a building characterized by a significant loss of structural features; and
- *Major damage*, which corresponds to Grade 5 in Figure 5-7, that is, the complete collapse of a building characterized by a loss of image features.

5.4.2 Problem Statement

Suppose that N pairs of bitemporal images of individual buildings, which are extracted from a pair of large-scale satellite images of an urban area stricken by a natural disaster, are provided. For an arbitrary pair of images, $\boldsymbol{u}0$ is used to denote the pre-event building image and $\boldsymbol{u}1$ for its postevent counterpart. Therefore, the imagery data set for conducting an object-based structural damage identification is expressed as follows:

$$\mathcal{D} = \left\{ (\boldsymbol{u}0, \boldsymbol{u}1)_n \mid \boldsymbol{u}0, \boldsymbol{u}1 \in \mathcal{R}^{D_n}; n = 1, 2, \ldots, N \right\} \tag{5-1}$$

where D_n is the dimension of the nth bitemporal image pair. To denote the categorical damage levels as discussed previously, an integer variable c is used:

- "$c=1$"—Minor damage,
- "$c=2$"—Moderate damage, and
- "$c=3$"—Major damage.

For a more compact notation, we denote a damage-level set as $C = \{c \mid c = 1, 2, 3\}$.

Therefore, the problem can be defined as follows. Given a bitemporal data set ($\boldsymbol{u}0$, $\boldsymbol{u}1$), which contains an engineering building captured before and after a disastrous event, a measure p_c is sought, which quantifies the *probability* that the building, defined by ($\boldsymbol{u}0$, $\boldsymbol{u}1$), has a structural damage level $c \in C$. To generate such a measure, the following parametric structure is formulated for p_c:

$$p_c = \psi\{\phi(\boldsymbol{u}0, \boldsymbol{u}1), c\} \tag{5-2}$$

The damage level for the building c^* is decided by choosing a damage level that maximizes p_c:

$$c^* = \underset{c}{\operatorname{argmax}}\, \psi\{\phi(\boldsymbol{u}0, \boldsymbol{u}1), c\} \tag{5-3}$$

Equations (5-1) to (5-3) define a generic mathematical framework for the process of feature extraction and damage classification. Particularly, Equation (5-2) represents a model that includes the following: (1) $\phi(\boldsymbol{u}0, \boldsymbol{u}1)$ represents a feature extractor that converts a bitemporal image pair (that is high-dimensional) to a low-dimensional and discriminative feature vector; (2) the notation $\psi(\cdot, c)$ represents a subsequent classification step that converts $\phi(\cdot, \cdot)$ to a probability measure that defines the underlying structural damage level between $\boldsymbol{u}0$ and $\boldsymbol{u}1$, if the damage level is c. Equation (5-3) simply states that the damage label is assigned by picking the label that returns the maximal probability measure by evaluating all possible damage labels.

5.4.3 Uncertainty in Remote Sensing Data

Equation (5-3) takes a probabilistic approach to damage classification; that is, a label that is predicted in principle has to be probabilistically characterized. This treatment accommodates the fact that RS data contain different sources of uncertainties when expressing damage levels. Two categories of uncertainties are involved in RS data collection and analysis.

1. *Data uncertainty*. The uncertainties in RS data collection (termed *data uncertainty*) include the stochastic ones owing to the instrument's electromagnetic and thermal and atmospheric noises. Epistemic data uncertainties are found in terms of incompleteness (e.g., owing to occlusion), errors (e.g., owing to misuse in sensor operation), and conflict (e.g., owing to the use of multiple types of sensors). Such epistemic imprecision becomes much significant when the emerging RS technologies are employed because of human operators (engineers, pilots, crowd, and so on). It is noted that such operator-caused uncertainties should be attributed to data uncertainties (operators are not users of RS data).
2. *User uncertainty*. Significant uncertainties subsequently arise in RS data interpretation introduced by the users (termed *user uncertainty*),

which is primarily epistemic. Traditionally, human-based visual analysis has been the primary cause of user uncertainty when visually assessing losses or hazard directly based on RS data. Recent crowd-based human analysis mitigates the level of such uncertainty owing to the underlying voting mechanism. This human-based analysis (such as damage labeling) is essential to create a training database for advanced machine learning–based data interpretation methods, the best of which needs to handle both data and user uncertainties.

5.4.4 Machine Learning–Based Damage Classification

The prediction model in Equation (5-3) generates a categorical damage prediction by seeking a class label $c \in \mathcal{C}$ that maximizes the function $p_c = \psi(\phi, c)$. It is worth noting that given a feature variable ϕ, the probability measure, p_c, is usually placed in the formalism of Bayesian decision theory and is interpreted as a posterior probability (Duda et al. 2000). In terms of the

$$p_c = P(c \mid \phi) = \frac{p(\phi \mid c)P(c)}{p(\phi)} \tag{5-4}$$

Equation (5-4) offers a principled way of modeling $P(c \mid \phi)$. This can be defined as a supervised pattern classification problem. The term *supervised* comes from the fact that the likelihood model $p(\phi \mid c)$ and the prior $P(c)$ in Equation (5-4) are not known from a priori knowledge. Instead, they are usually learned from data—in the context of pattern classification, which is termed *training data*:

$$\mathcal{D}^{tr} = \left\{ (\boldsymbol{u0}_n, \boldsymbol{u1}_n, c_n) \mid \boldsymbol{u0}_n, \boldsymbol{u1}_n \in \mathcal{R}^{D_n}, c_n \in \mathcal{C}; n = 1, 2, \ldots, N^{tr} \right\} \tag{5-5}$$

Given the training data set $\mathcal{D}^{tr}$, if an ideal damage feature extractor is obtained (i.e., for each input image pair, a low-dimensional damage feature vector $\phi_n = \phi[(\boldsymbol{u}0, \boldsymbol{u}1)n]$ is obtained), the construction of a probabilistic classifier according to Equation (5-4) is straightforward. First, a parametric structure for the likelihood functions $p(\phi \mid c)$'s and the prior probability, $P(c)$, can be estimated, and then the classification becomes ready through Equation (5-3). The integral performance of damage classification is determined by both the damage feature extraction step and the subsequent classification design. In general, it is critical to improve the performance of the damage feature extractor; then, an appropriate classifier should be sought.

5.4.5 Crowdsourcing-Based Damage Interpretation

To achieve the training data set in Equation (5-5), usually a small number of trained experts are recruited to inspect the remote sensing images and manually label the objects of interest. Such results give rise to the so-called visual ground truth (VGT), as mentioned by Chen and Hutchinson (2010). Subsequently, the computer-based machine learning method aims to compete with the human intelligence embedded in the VGT-based training data.

This practice implies that if this VGT approach is taken to a larger scale, a potentially highly promising computing venue—crowdsourcing-based computing or human-based computing—is obtained. Basically, if a large number of relatively trained human agents are "hired" or volunteer to conduct the image interpretation, this may bring in a significant performance accuracy gain and efficiency. After all, the machine learning methods presented previously indeed aim to compete against human intelligence, yet usually resulting in inferior performance in accuracy and generalization.

Crowdsourcing-based computing has been intensively studied in Computer Science, as summarized in a recent handbook (Michelucci 2013). Notably, in this handbook, Meier (2013) discussed the use of human computing for disaster response, including human-computing-based satellite data interpretation, and the involved challenges, benefits, and outcomes. In this chapter, the authors present one case study that demonstrates the successful use of crowdsourcing for RS-based loss assessment.

5.4.6 Emerging Big Data and Artificial Intelligence Methods

Disaster scenes are complex. As previously mentioned, with human-based VGT data as the training data, most machine learning algorithms aim to race against human intelligence toward providing a rapid and automated RS-based damage estimation approach. This normal may be disrupted by the recent advances in big data and artificial intelligence (AI), particularly the development of deep learning technology. With the RS big data and large-scale human-based labeling results, one expects that advanced AI methods may asymptotically approach the accuracy and robustness level of human intelligence.

The potential first comes from the recent development in big and labeled RS data at the object level by several private and government agencies. Three significant databases are recognized herein. The first is the recent launch of the NGA's XView program (Lam et al. 2018), which provides manually delineated more than 1 million bounding boxes in pre- and postevent RT images. The other large-scale remote sensing–based database

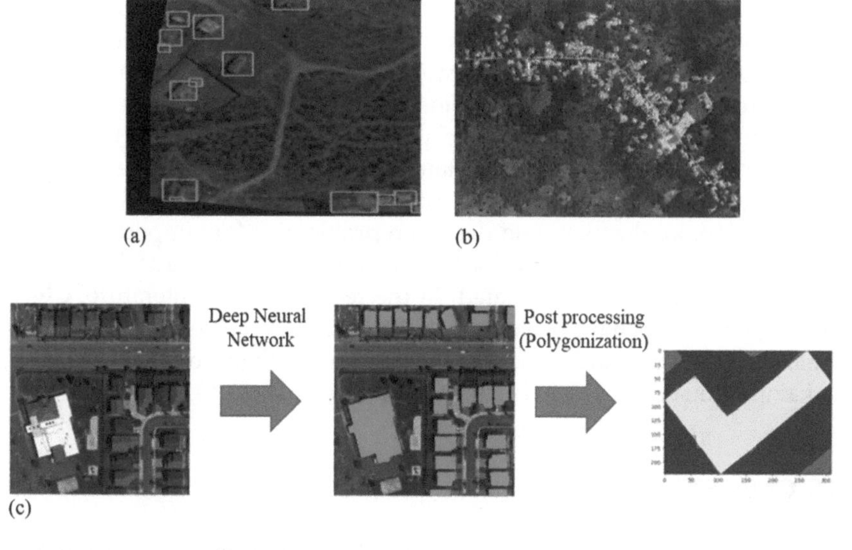

Figure 5-9. Existing "big" RS data with labels for built objects: (a) Sample training data in the NGA's Xview including ground-truth bounding boxes in the magnitude of 10^6, (b) sample image in the Maxar's Open Data Program that provides event-based footprints of built objects for significant disaster events since 2010, (c) Microsoft deep learning–based framework for building footprints, which provided more than 125 million building footprints in the United States.
Source: (a) Lam et al. (2018), (b) Maxar (2021), (c) recreated based on Microsoft (2018).

has been provided by Maxar through its Open Data Program (Maxar 2021) for major disasters since 2010. With their VGT nature, these two databases provide a tremendous boost for constructing an object-based high-resolution training/testing advanced AI-based damage detection algorithm. Equally significant is the Microsoft development of a deep learning–based building-footprint extraction framework, which already released an opensource data set for 125 million buildings in the United States (Microsoft 2018). Figure 5-9(a) and (b) illustrates the samples of the two disaster-event and object-based databases, and (c) illustrates the Microsoft framework for building extraction.

On the contrary, enabled by big data in different sectors, deep learning (DL) has reportedly outperformed many traditional machine learning methods, including the use of geospatial remote sensing data (LeCun et al. 2015, Zhang et al. 2016). A DL model first realizes the high-level abstraction by using a deep graph with multiple processing layers containing multiple linear or nonlinear transformations. It can then learn the intricate patterns in high-dimensional data (e.g., images) by breaking the complicated mapping into a series of simple mappings. From this perspective and especially for a supervised task, a DL model has an end-to-end workflow, in which the step of feature extraction and classification are both learned from the data. Therefore, a DL model largely avoids the subjective nature and burden of tuning both feature extraction and classification models as in the traditional machine learning paradigm. To date, the convolutional neural network (CNN)–based methods boast state-of-the-art performance in many object classification, localization, and detection problems. More important, many convolutional neural networks (CNNs) and their variants have been developed, trained, and validated. In the recent research literature, CNN-based DL methods have been used for RS data–based object detection (e.g., building detection), and some focused on damage detection (Bai et al. 2017, Vakalopoulou et al. 2015, Xu et al. 2018, Zhang et al. 2016).

5.5 CASE STUDIES

This section will focus on several case studies in previous disasters that employed RS data successfully for disaster-induced damage detection and mapping.

5.5.1 E-DECIDER's Rapid Remote Sensing–Based Damage Assessment

E-DECIDER was a NASA-funded project that aimed to develop computing infrastructure and remote sensing–based products for rapid disaster response and decision-making tools for the first responders (NASA 2019). The first-decision support product that the E-DECIDER team developed was used for the 2011 Tohoku Japan earthquake and tsunami. The product was directly delivered to the Japanese government for response efforts through the United States and Japanese International Charter representatives. Immediately following the earthquake, the Japanese government provided a list of cities along the northwestern coast of Honshu that was known to have sustained significant damage from the tsunami and cities for which they had no contact since the tsunami came ashore. The USGS provided access to all data collected for the earthquake and tsunami response, including restricted imagery. Digital change detection results showing tsunami inundation using MODIS imagery are presented in Figure 5-10 (with red pixels indicating areas of inundation extent).

5.5.2 Machine Learning–Based Building Damage Mapping

In this case study, we emphasize an important aspect of disaster assessment when using RS data and machine learning methods—to deal with the inherent uncertainties in RT data and the ensuing training data,

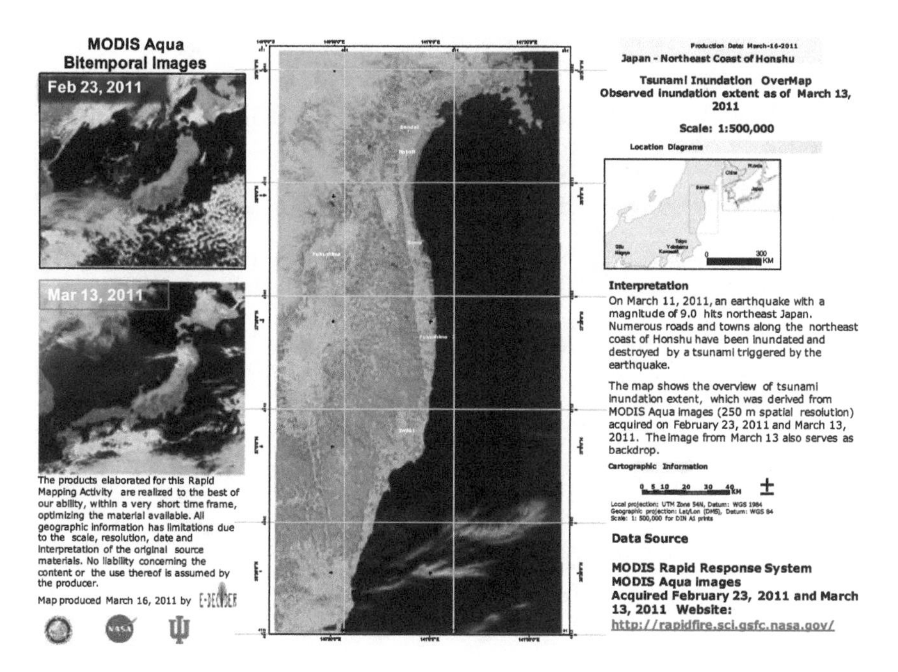

Figure 5-10. Remote sensing–based digital change detection of the northern Honshu coast using MODIS imagery in responding to the 2011 Jana Earthquake and Tsunami. Red pixels indicate tsunami inundation extent identified by E-DECIDER.

probabilistic methods are desired. For this purpose, a pair of georeferenced, pan-sharpened, bitemporal images captured by the Quickbird satellite (Digital Globe 2011) with a resolution of 60 cm/pixel were used (Figure 5-3): The pre-event image was captured before the India Ocean Tsunami on June 23, 2004, and the postevent image was captured two days after the tsunami, on December 28, 2004. Postdisaster reconnaissance reports (World Bank 2005) indicate that approximately 1.3 million structures were damaged because of the devastating effects of the tsunami within the area of Aceh Province of Indonesia alone. Figure 5-11(a) show only a portion of the Banda Aceh area in the Aceh Province of Indonesia in a set of bitemporal image patches.

Two independent student analysts performed manual building extraction and object-based damage classification, providing RT-based visual-inspection ground-truth (VGT) data for the probabilistic learning-based classification and validation. The VGT results are illustrated in Figure 5-11(b, left), in which a total of 463 structures are rendered in different colors, with 268 (57%) structures showing major damage (red), 91 (20%) showing moderate damage (yellow), and 104 (22%) showing minor damage (green). The VGT map in Figure 5-11(b) may serve as a direct

Figure 5-11. (a) Portion of bitemporal images of Banda Aceh, Indonesia region, which was taken (left) before and (right) after the 2004 Indian Ocean Tsunami, (b) training/validation input data (left), predictive damage map (right) (red: major damage, yellow: moderate damage, green: minor or no damage).
Source: Based on resolution-downgraded images licensed through PDC (2008).

product for the end users (disaster responders). However, such visual inspection–based assessment is time-consuming (given in the GT report that millions of buildings were damaged). Furthermore, when multiple persons are involved in the process, subjective inconsistencies will arise. Therefore, we explored the possibility of conducting supervised damage classification using a small portion of VGT results agreed upon by all visual analysts. In Figure 5-11(b, left), the building objects confined in the upper-right bounding box were used to illustrate this process.

We proposed a probabilistic damage classification method, which addressed a major limitation when adopting a deterministic change detection or machine learning method., ,that is, by applying a probabilistic approach, the classification of structural damage provides posterior probabilities, which can quantify the decision uncertainties and be further used to obtain regional urban damage classification (e.g., at a city block level). For a detailed formulation of this probabilistic classifier and its application to multiclass and regional damage classifications, readers may refer to Chen and Hutchinson (2010). The trained multiclass damage classifier enables the prediction of damage levels for the remaining urban buildings. Figure 5-11(b, right) displays the predicted result. A quantitative evaluation indicates that the overall classification accuracy is about 78%. In particular, for buildings with minor or major damage, the recall and precision accuracies are both ≥80%. However, for buildings with moderate damage, the recall and precision accuracies are <40%. This relatively low performance in classifying moderate damage is also observed during the visual identification process. It arises because the determination of moderate damage involves more subjective uncertainties compared with the other two (extreme) damage cases. Overall, one can observe from the early practice that when small samples of training data are used, given the underlying uncertainties (RT data intrinsic and the human-related), machine learning–based accuracy is not comparable to human intelligence.

5.5.3 Crowdsourcing-Based Damage Interpretation and Assessment

This case study accounts for how crowdsourcing is used in RT-based damage assessment (Corbane et al. 2011, Ghosh et al. 2011, Bevington et al. 2015). The event was the 2010 Haiti earthquake, which, in less than a minute, leveled approximately 20% of the buildings in the greater Port-au-Prince (PaP) city; killed close to a quarter of a million people; injured as many; and left a million individuals homeless. The visual interpretation of damage in Haiti started with Phase 1 using satellite data. These data proved to be extremely valuable in determining the scope of the disaster and in prioritizing both aerial and field surveys. The "counting" of the number of severely damaged buildings using satellite images (Phase 1) was relatively quick. However, when compared with the higher-resolution aerial images (Phase 2), these counts were found to be about a quarter of what was assessed using the aerial data. Some of the reasons for the underestimation included difficulty in discerning damage using this relatively low-resolution imagery, the damage being so prevalent in many of the areas that analysts were not able to distinguish damage to individual buildings.

The ImageCat team working with the World Bank and several professional engineering societies, including the Earthquake Engineering Research Institute, launched a novel effort to use crowdsourcing or human computing as a tool to rapidly assess earthquake damage in the greater PaP region. Higher-resolution aerial images at the resolution of 15 cm/pixel were used, which allowed analysts to "see" damage with such precision that damage that usually is difficult to see (e.g., partial roof collapses, shifting of buildings off foundations, and so on). This crowdsourcing effort—called GEO-CAN (Global Earth Observation—Catastrophe Assessment Network) involved over 600 earthquake experts representing 23 countries from 131 private, government, and academic institutions who dedicated at least several hours each in helping to assess damage in this event. See Ghosh et al. (2011) for more logistic details on this crowdsourcing effort.

During Phase 2, close to 30,000 buildings in PaP and surrounding areas were identified by the GEO-CAN community as having either Grade 4 (very heavy damage) or Grade 5 (destroyed), according to the European Macroseismic Scale (Figure 5-8). This represents roughly 10% of the total building stock in the affected area. Table 5-3 summarizes the results of these organized crowdsourcing efforts, and spatial lessons were learned as noted in the following:

- As part of the Phase-2 damage assessment, it became very evident that a significant portion of the damage count was being omitted because the type of damage that was occurring was not visible in the aerial images. To address this deficiency, the ImageCat team initiated two major activities: fieldwork to develop damage distributions that could serve as the basis for extrapolating aerial results to other lower damage states, and evaluation of oblique imagery (Pictometry data) to supplement the field results in areas that were physically hard to access or were in remote locations. Using the field and Pictometry data, we estimated that the Grade 4 and 5 building counts were underestimating the amount of building damage by at least a factor of two in some cases.
- A series of damage distributions (i.e., tables) were produced that were eventually used by the joint PDNA team to extrapolate the aerial damage results to the full damage total. As a result of this joint analysis, close to 300,000 buildings were analyzed, with approximately 20% of these buildings in the destroyed or heavily damaged categories.
- One of the essential products from the GEO-CAN analysis was the delineation of building footprints for buildings experiencing at least Grade 4 damage. This information was crucial in quantifying the amount of building floor space that was eliminated by the earthquake. By combining Pictometry and field data, the total amount of floor area associated with all damaged buildings (as determined by the joint PDNA damage assessment) was estimated at about 41 million square meters, of which 18% is associated with Grade 4 and 5 buildings.

5.6 CONCLUSIONS AND RECOMMENDATION FOR PRACTICE

5.6.1 Conclusions and Remarks

Through reviewing both traditional and emerging remote sensing (RT) technologies, we state that remote sensing technologies can provide rapid means to obtain valuable information for both first responders and the public. With a focus on optical remote sensing data, and through

Table 5-3. Summary of Damage and Replacement Statistics Generated by the ImageCat/GEO-CAN Damage Assessment.

	Number of buildings					
	Damage grade (EMS 98)					
Building category	Grade 1: negligible/no visible damage	Grade 2: moderate damage	Grade 3: substantial damage	Grade 4: very heavy damage	Grade 5: destroyed	Total
Residential (low density)	65,259	5,675	10,404	4,389	8,852	**94,579**
Residential (high density)	14,909	1,683	2,645	1,382	3,187	**23,807**
Commercial	283	1,039	992	646	1,763	**4,724**
Industrial	71	259	247	253	348	**1,178**
Downtown	107	393	376	380	532	**1,788**
Informal/shanty	15,289	485	2,427	1,921	4,146	**24,268**
Agricultural	6,131	533	977	329	915	**8,886**
Open land	64	6	10	4	9	**93**
Total	**102,114**	**10,074**	**18,078**	**9,304**	**19,752**	**159,322**
	Area (square meters)					
Building Category	Grade 1	Grade 2	Grade 3	Grade 4	Grade 5	**Total**
Residential (low density)	10,702,511	930,653	1,706,197	719,796	1,451,728	**15,510,886**
Residential (high density)	1,971,019	222,534	349,697	182,700	421,321	**3,147,271**

(Continued)

Table 5-3. (*Continued*)

	Number of buildings					
	Damage grade (EMS 98)					
Building category	Grade 1: negligible/no visible damage	Grade 2: moderate damage	Grade 3: substantial damage	Grade 4: very heavy damage	Grade 5: destroyed	Total
Commercial	44,212	162,112	154,743	100,776	275,028	**736,871**
Industrial	9,460	34,688	33,112	33,851	46,562	**157,674**
Downtown	18,616	68,257	65,154	65,930	92,302	**310,259**
Informal/shanty	966,255	30,675	153,374	121,407	262,027	**1,533,738**
Agricultural	503,980	43,824	80,345	27,044	75,213	**730,406**
Open land	5,267	458	840	329	740	**7,633**
Total	**14,221,319**	**1,4931,201**	**2,5431,461**	**1,2511,834**	**2,6241,922**	**22,134,737**
	Replacement or repair cost (US$)					
Cost (US$/m^2)	40	100	300	500	500	**Total**
Total cost (millions)	**569**	**149**	**763**	**626**	**1,312**	**3,420**

Source: Ghosh et al. (2011).

demonstrating a few case-based studies, this chapter further demonstrates the effectiveness of RS-based damage detection.

We acknowledge that remote sensing data have become more abundant, and derived products are more commonly used today. They are collected not only through the traditional satellite or airborne sensors but also by personalized aerial or mobile platforms (e.g., UAVs or smartphones). These data sets, if all utilized, will lead to the notion of disaster-scene big data. The recent databases and deep learning–based AI technologies hold substantial promise that pushes the boundary of RT-based automatic damage detection and mapping to the next level. One may expect that this trend may asymptotically approach the accuracy of human intelligence–based building extraction, damage detection, and mapping.

We foresee that remote sensing technologies, including various RS platforms, imaging sensors, social networks, mobile-cloud computing, Internet of Things, and AI-enabled analytics, will be highly integrated and automated in the future. They will continue strengthening their significant roles in building resourcefulness and rapidness into community resilience by providing critical disaster and hazard analytics data in real time and will be integrated with other local resilience monitoring technologies toward enhancing the robustness and redundancy in both community and asset resilience. Yet, the future RS as a whole and as a highly comprehensive and robust utility will itself become resilient when dealing with disasters or any disruption, if such integration becomes a reality. The authors envision that the RS-based service will become a transparent service to anybody at any time and at any location for emergency-triggered decision making.

5.6.2 Recommendation for Practice

With these technical contents introduced previously, and considering both traditional and emerging RS technologies to this date, it is acknowledged that derived products through computational processing and modeling of remote-sensing data have yet to be commonly used in practice. Often available is a certain level of GIS readiness added to the remote sensing data, and visual interpretation is still indispensable. Given these facts, the authors suggest and recommend the following practices, which can be readily achieved today by most agencies and by the general body of first responders, including emergency coordinators, responders, analysts, and field inspectors.

1. Remote sensing data and derived products should be achieved in a GIS-ready system that provides remote and mobile access. These archived data can provide predisaster preparedness or immediate postdisaster or during-disaster reference when the postdisaster data are not available. Taking advantage of modern centralized cloud

infrastructure that provides distributed and real-time services for mobile users has become a new normal and should be implemented in practice.

2. Attention needs to be paid when remote sensing data come with different timelines and different resolutions. In the immediate aftermath or during a disaster event, it is usually the moderate-resolution data (e.g., Landsat data) that become available first, then other free-access high-resolution data (e.g., Sentinel-1/2 data), and finally high-resolution to very-high-resolution commercial (or ad hoc access) data. End users of these data need to make decisions based on these practical limitations of their different timeliness and resolution. In general, low- and moderate-resolution data can provide a reference of the hazard coverage and extent about an unfolding disaster; and high-resolution data provide both hazard intensity and infrastructure and structure damage information.
3. Emerging crowdsourcing technologies based on the use of social networks and personal or community-based UAVs can provide important, agile, and detailed disaster data. However, these data sets are often much less structured and possess more uncertainties. Organized social-network approaches or the use of UAVs are preferred, which lead to more of a citizen-scientist approach providing more structured and trustful data. Nonetheless, end users are responsible for developing standard protocols for employing these technologies (particularly, for the use of UAVs or drones, they are subject to the regulations of FAA and local airspace classifications).
4. Other geospatial tools, including predictive loss modeling and forecast of potential hazards, are well developed, and some of them have become standard methodologies (e.g., the use of HAZUS-MH for loss estimation for different hazards). If no remote sensing data are available in the aftermath of a disaster, these tools serve as the most substantial source of information for understanding the hazard intensity and the potential losses. If RT-based products exist, because of their nature of being real measurements of structural changes related to buildings and civil infrastructure, they provide more trustful damage information.
5. Advanced machine vision or artificial intelligence (AI) methods have not been implemented to this date as a service, which is being considered by several agencies and research efforts. Active collaboration among the end users, research and development personnel, and commercial entities are called for and expected to deliver service-based infrastructure in the near future.
6. Communication of either predictive, monitoring-based, or any remote sensing–based derived products needs to be conducted with a risk-based framework. However, significant research in developing

standard methodologies is in great need. It is believed that before the establishment of these standard methodologies, empirical protocols and in-situ discretion based on past experience are probably the best mechanisms for civil infrastructure stakeholders.

ACKNOWLEDGMENTS

Dr. Margaret Glasscoe is supported by the Jet Propulsion Laboratory, California Institute of Technology, under a contract with the National Aeronautics and Space Administration (80NM0018D0004). Dr. Bandana Kar is supported by UT-Battelle, LLC, under Contract No. DE-AC05-00OR22725 with the US Department of Energy (DOE).

REFERENCES

Bai, Y., C. Gao, S. Singh, M. Koch, B. Adriano, E. Mas, et al. 2017. "A framework of rapid regional tsunami damage recognition from post-event TerraSAR-X imagery using deep neural networks." *IEEE Geosci. Remote Sens. Lett.* 15 (1): 43–47.

Bevington, J., R. Eguchi, S. Gill, S. Ghosh, and C. Huyck. 2015. "A comprehensive analysis of building damage in the 2010 Haiti earthquake using high-resolution imagery and crowdsourcing." In *Time-sensitive remote sensing*, L. D. Christopher, S. A. Douglas, and C. L. Lyoyd, eds., 131–145. New York: Springer.

Bishop, C. M. 2006. *Pattern recognition and machine learning*. New York: Springer.

Bruneau, M., S. E. Chang, R. T. Eguchi, G. C. Lee, T. D. O'Rourke, A. M. Reinhorn, et al. 2003. "A framework to quantitatively assess and enhance the seismic resilience of communities." *Earthquake Spectra* 19 (4): 733–752.

Chan, T. F., and J. J. Shen. 2005. *Image processing and analysis: Variational, PDE, wavelet, and stochastic methods*. University City, PA: SIAM.

Chen, J., Z. Chen, and C. Beard. 2018. "Experimental investigation of aerial–ground network communication towards geospatially large-scale structural health monitoring." *J. Civ. Struct. Health Monit.* 8 (5): 823–832.

Chen, J., Z. Dai, and Z. Chen. 2019. "Development of radio-frequency sensor wake-up with unmanned aerial vehicles as an aerial gateway." *Sensors* 19 (5): 1047.

Chen, Z. 2014. "A micro-UAV approach to earthquake disaster sensing: A critical review." In *Proc., NCEE 2014—10th US National Conf. on*

Earthquake Engineering: Frontiers of Earthquake Engineering. Oakland, CA: Earthquake Engineering Research Institute.

Chen, Z., and T. C. Hutchinson. 2010. "Probabilistic urban structural damage classification using bitemporal satellite images." *Earthquake Spectra* 26: 87–109.

Cheng, N., W. Xu, W. Shi, Y. Zhou, N. Lu, H. Zhou, et al. 2018. "Air–ground integrated mobile edge networks: Architecture, challenges, and opportunities." *IEEE Commun. Mag.* 56 (8): 26–32.

Chiroiu, L. 2005. "Damage assessment of the 2003 Bam, Iran, earthquake using Ikonos imagery." *Earthquake Spectra* 21 (S1): S219–S224.

Corbane, C., D. Carrion, G. Lemoine, and M. Broglia. 2011. "Comparison of damage assessment maps derived from very high spatial resolution satellite and aerial imagery produced for the Haiti 2010 earthquake." *Earthquake Spectra* 27 (S1): S199–S218.

EMS (European Macroseismic Scale). 1998. "European macroseismic scale (EMS-98)," edited by Grünthal G. Walferdange. Luxembourg: European Centre for Geodynamics and Seismology.

Farrar, C. R., and K. Worden. 2006. "An introduction to structural health monitoring." *Philos. Trans. R. Soc. London, Ser. A* 365 (1851): 303–315.

FEMA. 2012. *Integrated Alert and Warning System (IPAWS)*. Washington, DC: FEMA.

Frangopol, D. M., and M. Soliman. 2016. "Life-cycle of structural systems: Recent achievements and future directions." *Struct. Infrastruct. Eng.* 12 (1): 1–20.

Frankel, A., C. Mueller, T. Barnhard, E. Leyendecker, R. Wesson, S. Harmsen, et al. 2000. "USGS national seismic hazard maps." *Earthquake Spectra* 16 (1): 1–19.

Ghobaraha, A., M. Saatcioglub, and I. Nistorb. 2006. "The impact of the 26 December 2004 earthquake and tsunami on structures." *Eng. Struct.* 28 (2): 312–326.

Ghosh, S., C. K. Huyck, M. Greene, S. P. Gill, J. Bevington, W. Svekla, et al. 2011. "Crowdsourcing for rapid damage assessment: The Global Earth Observation Catastrophe Assessment Network (GEO-CAN)." *Earthquake Spectra* 27 (S1): S179–S198.

Gladwin, H., J. K. Lazo, B. H. Morrow, W. G. Peacock, and H. E. Willoughby. 2009. "Social science research needs for the hurricane forecast and warning system." *Bull. Am. Meteorol. Soc.* 90 (1): 25–29.

Glaessgen, E., and D. Stargel. 2012. "The digital twin paradigm for future NASA and US Air Force vehicles." In *Proc., 53rd AIAA/ASME/ASCE/AHS/ASC Structures, Structural Dynamics and Materials Conf., 20th AIAA/ASME/AHS Adaptive Structures Conf., 14th AIAA*. Published online. URL: https://doi.org/10.2514/6.2012-1818. Accessed September 1, 2021.

Grabill, J. T., and W. M. Simmons. 1998. "Toward a critical rhetoric of risk communication: Producing citizens and the role of technical communicators." *Tech. Commun. Q.* 7 (4): 415–441.

Gusella, L., B. Adams, G. Bitelli, C. Huyck, and A. Mognol. 2005. "Object-oriented image understanding and post-earthquake damage assessment for the 2003 Bam, Iran, earthquake." *Earthquake Spectra* 21 (S1): S225–S238.

JBEC (Japan Bridge Engineering Center). 2011. *Damage to highway bridges caused by the 2011 Tohoku-Oki earthquake*. Tokyo: JBEC.

JSCE (Japanese Society of Civil Engineers). 2005. *The damage induced by Sumatra earthquake and associated tsunami of December 26, 2004*. Tokyo: JSCE.

Kar, B. 2018. "Results of an integrated approach to geo-target at-risk communities and deploy effective crisis communication approaches." In *Emergency alert and warning systems: Current knowledge and future research directions*, 70–73. Washington, DC: National Academies Press.

Kar, B., D. Cochran, X. Liu, J. Zale, J. Dickens, and N. Callais. 2016. *An integrated approach to geo-target at-risk communities and deploy effective crisis communication approaches*. Washington, DC: Dept. of Homeland Security-Science and Technology Directorate.

Kohiyama, M., and F. Yamazaki. 2005. "Damage detection for 2003 Bam, Iran, earthquake using Terra-Aster satellite imagery." *Earthquake Spectra* 21 (S1): S267–S274.

Krimsky, S. 2007. "Risk communication in the internet age: The rise of disorganized skepticism." *Environ. Hazards* 7 (2): 157–164.

Lam, D., R. Kuzma, K. McGee, S. Dooley, M. Laielli, M. Klaric, et al. 2018. "xView: Objects in context in overhead imagery." Preprint. http://arXiv.org/abs/1802.07856.

LeCun, Y., Y. Bengio, and G. Hinton. 2015. "Deep learning." *Nature* 521 (7553): 436.

Lillesand, T., R. W. Kiefer, and J. Chipman. 2015. *Remote sensing and image interpretation*. Hoboken, NJ: Wiley.

Liu, X., B. Kar, C. Zhang, and D. M. Cochran. 2018. "Assessing relevance of tweets for risk communication." *Int. J. Digital Earth* 12 (7): 781–801.

Maxar. 2021. "Open data program." Accessed May 1, 2021. https://www.maxar.com/open-data.

Meier, P. 2013. "Human computation for disaster response." In *Handbook of human computation*, P. Michelucci, ed., 95–104. New York: Springer.

Michelucci, P. 2013. *Handbook of human computation*. New York: Springer.

Microsoft. 2018. "Microsoft releases 125 million building footprints in the US as open data." Accessed May 1, 2021. https://blogs.bing.com/maps/2018-06/microsoft-releases-125-million-building-footprints-in-the-us-as-open-data.

Mileti, D. S., and L. Peek. 2000. "The social psychology of public response to warnings of a nuclear power plant accident." *J. Hazard. Mater.* 75 (2): 181–194.

Murphy, R. R., E. Steimle, C. Griffin, C. Cullins, M. Hall, and K. Pratt. 2008. "Cooperative use of unmanned sea surface and micro aerial vehicles at Hurricane Wilma." *J. Field Rob.* 25 (3): 164–180.

NASA (National Aeronautics and Space Administration). 2019. "Earthquake data enhanced cyber-infrastructure for disaster evaluation and response (E-DECIDER)" Accessed September 1, 2021. https://appliedsciences.nasa.gov/what-we-do/projects/earthquake-data-enhanced-cyber-infrastructure-disaster-evaluation-and-response.

NRC (National Research Council). 2012. *Disaster resilience: A national imperative.* Washington, DC: National Academies Press. Pix4D. 2021. Pix4D: Photogrammetry software for professional drone mapping. Accessed September 1, 2021. https://www.pix4d.com/.

PDC (Pacific Disaster Center). 2008. "PDC Global." Accessed September 1, 2021. http://www.pdc.org.

Pratt, K. S., R. Murphy, S. Stover, and C. Griffin. 2009. "CONOPS and autonomy recommendations for VTOL small unmanned aerial system based on Hurricane Katrina operations." *J. Field Rob.* 26 (8): 636–650.

Qiuchen Lu, V., A. K. Parlikad, P. Woodall, G. D. Ranasinghe, and J. Heaton. 2019. "Developing a dynamic digital twin at a building level: Using Cambridge Campus as case study." In *Proc., Int. Conf. on Smart Infrastructure and Construction 2019: Driving Data-Informed Decision-Making*, 67–75. London: Institute of Civil Engineers.

Rathje, E., and M. Crawford. 2004. "Using high resolution satellite imagery to detect damage from the 2003 Northern Algeria Earthquake." In *Proc., 13th World Conf. on Earthquake Engineering, Vancouver, British Columbia.* Vancouver: Canadian Association for Earthquake Engineering and International Association for Earthquake Engineering.

Rees, W. G. 2013. *Physical principles of remote sensing.* Cambridge, UK: Cambridge University Press.

Rejaie, A., and M. Shinozuka. 2004. "Reconnaissance of Golcuk 1999 earthquake damage using satellite images." *J. Aerosp. Eng.* 17 (1): 20–25.

Schneider, P. J., and B. A. Schauer. 2006. "Hazus—Its development and its future." *Nat. Hazard. Rev.* 7 (2): 40–44.

Shawe-Taylor, J., and N. Cristianini. 2004. *Kernel methods for pattern analysis.* Cambridge, UK: Cambridge University Press.

Shim, C.-S., N.-S. Dang, S. Lon, and C.-H. Jeon. 2019. "Development of a bridge maintenance system for prestressed concrete bridges using 3D digital twin model." *Struct. Infrastruct. Eng.* 15 (1): 1–14.

SIC (Satellite Imaging Corporation). 2021. "WorldView-3 Satellite Sensor (0.31m)." Accessed September 1, 2021. https://www.satimagingcorp.com/satellite-sensors/worldview-3/.

UN/ISDR (United Nations/International Strategy for Disaster Reduction). 2004. *Living with risk: A global review of disaster reduction initiatives*. New York: UN/ISDR.

UNDRR (United Nations Office for Disaster Risk Reduction). 2015. *Sendai Framework for Disaster Risk Reduction 2015–2030*. Geneva: UNDRR.

USGS (US Geological Survey). 2008. *The 26 December 2004 Indian Ocean Tsunami: Initial Findings from Sumatra*. Reston, VA: USGS. Accessed September, 1, 2021. https://cmgds.marine.usgs.gov/data/sumatra05/index.html.

Vakalopoulou, M., K. Karantzalos, N. Komodakis, and N. Paragios. 2015. "Building detection in very high resolution multispectral data with deep learning features." In *Proc., 2015 IEEE Int. Geoscience and Remote Sensing Symp.*, 1873–1876. New York: IEEE.

Walsh, K. J., S. J. Camargo, G. A. Vecchi, A. S. Daloz, J. Elsner, K. Emanuel, et al. 2015. "Hurricanes and climate: The US CLIVAR working group on hurricanes." *Bull. Am. Meteorol. Soc.* 96 (6): 997–1017.

World Bank. 2005. *Indonesia: Preliminary damage and loss assessment, the December 26, 2004 natural disaster*. Washington, DC: World Bank.

World Bank. 2011. "2010 Haiti earthquake: Post-disaster building damage assessment using satellite and aerial imagery interpretation, field verification and modeling techniques." Global Facility for Disaster Reduction and Recovery (GFDRR). Accessed May 1, 2021. https://www.gfdrr.org/sites/default/files/publication/2010haitiearthquakepost-disasterbuildingdamageassessment.pdf.

Xu, Y., L. Wu, Z. Xie, and Z. Chen. 2018. "Building extraction in very high resolution remote sensing imagery using deep learning and guided filters." *Remote Sens.* 10 (1): 144.

Yamazaki, F., Y. Yano, and M. Matsuoka. 2005. "Visual damage interpretation of buildings following the 2003 Bam, Iran, earthquake." *Earthquake Spectra* 21 (S1): S329–S336.

Yeum, C. M., and S. J. Dyke. 2015. "Vision-based automated crack detection for bridge inspection." *Comput.-Aided Civ. Infrastruct. Eng.* 30 (10): 759–770.

Yoon, H., V. Hoskere, J.-W. Park, and B. Spencer. 2017. "Cross-correlation-based structural system identification using unmanned aerial vehicles." *Sensors* 17 (9): 2075.

Zhang, L., L. Zhang, and B. Du. 2016. "Deep learning for remote sensing data: A technical tutorial on the state of the art." *IEEE Geosci. Remote Sens. Mag.* 4 (2): 22–40.

CHAPTER 6

ROLE OF SEISMIC ISOLATION AND PROTECTION DEVICES IN ENHANCING STRUCTURAL RESILIENCE

Fariborz M. Tehrani, Maryam Nazari, Ali Naghshineh

6.1 INTRODUCTION

6.1.1 Resilience in Earthquake-Resistant Design

Recent advances in the field of structural engineering have substantially shifted engineering paradigms in earthquake-resistant design. Although life safety remains the paramount goal in engineering design, the path to a safe design has been influenced by innovative means and materials of structural systems as well as robust computational and analytical capabilities in the areas of seismology, mathematical modeling, structural mechanics, health monitoring, and other relevant fields. As a result, the performance-based design (PBD) approach has gained popularity in recent decades. The performance-based earthquake engineering (PBEE) exceeds expectations of merely surviving and addresses structural performance at multiple levels, including operational, immediate occupancy, life safety, and collapse prevention (SEAOC 1995). The comprehensive approach of the PBEE framework involves four major stages of hazard, structural, damage, and loss analyses (Sullivan et al. 2014). These processes introduce uncertainty as an essential challenge in the assessment of structural response, applied loads, and expected damage. Thus, an estimation of structural vulnerability to extreme events and restoration of structural functionality after such events inherits those uncertainties (Kanno et al. 2017). Furthermore, as PBD changes the procedures and expectations of earthquake-resistant structural design, lessons learned from recent earthquakes have highlighted the importance of nonstructural components in the overall estimation of losses and damage. Nonstructural components contribute to a significant portion of total investments in buildings and, thus, may contribute to substantial financial losses during earthquakes (Filiatrault and Sullivan 2014). In addition, the shift in earthquake design and resultant structural performance during recent earthquakes have also surfaced other sources of loss, such as business disruption and economic decline (Mayes et al. 2012). Such limited tolerance for damage after major earthquake events has led structural engineers to adopt sustainable and resilient structural systems (Grigorian and Grigorian 2018). These pieces of evidence signify the need for considering resilience as the primary objective of earthquake-resistant design of structures (Gilmore 2012, Pessiki 2017, Mahin 2008, 2017). Key components of resilience, including robustness, resourcefulness, rapidity, and redundancy, interact with other structural characteristics, such as safety and reliability through PBD (Ettouney 2014). This chapter shows how seismic isolation practically enhances resilience by isolating and protecting structures from transferred loads during earthquake events and/or damping the energy of transferred motions to structures during such events.

6.1.2 Defining Resilience for Structures Subject to Seismic Events

A general definition for the seismic resilience of structures relates to the degradation of structural performance owing to an earthquake event. Several parameters may influence such degradation, including the state of the structure before the event, intensity of the event, and recovery process after the event (Vugrin et al. 2010, Bruneau et al. 2003). Structures are dynamic systems that tend to degrade throughout their service time span unless supported by maintenance, upgrade, and retrofit efforts to cope with their expected and unexpected variations of demand. Thus, the state of structural health before the occurrence of a seismic event will influence the intensity of degradation owing to that event, as well as susceptibility to further degradation owing to secondary risks triggered by the earthquake. Obviously, the vulnerability to earthquakes influences the degradation of the structure as well. Further, the efficiency of the recovery process determines the degradation of the structure. This implies the importance of a multi-hazard approach to ensure the resilience of structures, that is, considering more than one hazard and ideally all known hazards, and their interactions. Furthermore, the inherent uncertainties in the characteristics of future seismic events introduce the need to consider the worst-case scenario in structural resilience (Takewaki et al. 2003). Such consideration elevates design concerns for structures that are typically fit to perform after the most probable case and invites further discussions about the theory and practice of design for the worst-case event, from the definition of the worst case to the implementation of design within an economically constrained environment. Finally, the recovery process after the event, with respect to means (equipment), methods (techniques), and speed of recovery, influences the resilience of a structure. The uncertainty in the state of a structure immediately after a seismic event has roots in uncertainties in the state of structural health and the intensity of the earthquake. Moreover, the nature of the recovery process, disregarding the scale and magnitude, is characteristically different from planned maintenance and upgrade efforts owing to socioeconomic pressures that are common after natural disasters. Therefore, it is not uncommon for recovery projects to have a limited scope that will not restore the performance of a structure to the initial level. Thus, it is important for the definition of resilience to include total degradation in the performance of a structure after an earthquake and not just in the immediate aftermath of the event. This is the essence of the system's robustness.

6.1.3 Seismic Isolation

Seismic isolation has gained substantial attention in the last several decades in many countries and thus, has given seismically isolated structures multiple opportunities for observations and assessment during real earthquakes. Many new seismic isolation systems proposed in the past recent years have proved to be practical by establishing a satisfactory structural performance after the occurrence of earthquakes. The design approach in seismic isolation utilizes technologies that minimize the effects of earthquakes on structures and thus, reduce consequent damage and economic losses owing to downtime. The reliability of isolation systems has roots in their low sensitivity to structural and site characteristics. Such reliability promotes isolation systems in probabilistic approaches to the design of earthquake-resistant structures, such as PBD, and exemplifies them as mechanisms to enhance the resilience of structures, which directly impacts the cost of recovery after earthquakes. Furthermore, the preventive nature of the seismic isolation technique makes it a sustainable practice in respect to the conservation of resources, in addition to preservation of life and safety (Tehrani 1992, Naeim and Kelly 1999, Naeim 2001, Bozorgnia and Bertero 2004).

6.1.4 Design Philosophy

The two major concerns in the behavior of structural systems during strong ground motions are story drift and floor accelerations. The balance between these demands tends to categorize structures as rigid or flexible. These categories further define structural requirements and criteria to achieve proper ductility, detail connections of nonstructural components, and so on. Overall objectives of these structural criteria require structures to (1) remain operational after frequent lower-magnitude earthquakes without damage, (2) sustain no major damage in occasional earthquakes, and (3) survive rare higher magnitude earthquakes without collapse, as defined by the relevant codes and standards. The performance-based design approach covers these appropriations of structural behavior with the level of applied ground motions. These basic objectives provide guidelines to assure that the capacity of structures will be more than earthquake demands throughout the design ground-motion event in respect to forces and deformations (Mayes et al. 1988, Tehrani 1993a). This simple design principle is idealistic because the strength and the stiffness of common structures are proportional, and, thus, stiffer structures tend to absorb more forces because of their higher rigidities and lower natural periods. Consequently, standard practices in structural engineering guide designers to enhance the ductility of structural materials and components, which is not always economical or even feasible without sacrificing nonstructural systems (Ferritto 1991, Kelly 1990, Tehrani 1993a, 1994, Tehrani and Hassani 1994a). The concept of seismic isolation addresses this problem by isolating strong and rigid structural elements using flexible and ductile isolation devices (Buckle 1989, Buckle and Mayes 1990, Kelly 1990, Tehrani 1992).

6.1.5 Performance-Based Design

The concept of performance-based seismic design is increasingly being embraced in different jurisdictions. Displacement control and objective-based solutions are some aspects of performance-based codes provided by the National Building Code of Canada (NBCC 2015), Eurocode 8, and ASCE 7-16. Performance assessment is an essential feature of a performance-based design. The overall structural performance cannot be explicitly defined during a given seismic event by current seismic design codes (e.g., ASCE 7-16) (ASCE 2017), which are prescribed-based and focus on the strength and capacity of the structural members (Naghshineh et al. 2018). Performance-based design deviates from this in that it is based on the objectives with a specific level of structural performance during a certain level of a seismic event, as demonstrated in Figure 6-1 (base shear demand versus lateral deformation).

The performance-based seismic design enables the design team to work together to evaluate the appropriate levels of ground motion and performance objectives for the building and the nonstructural elements to meet the owner's expectations. Existing literature indicates that the seismic response during actual earthquakes may differ from the predicted amount during design procedures, as demonstrated in the following cases.

Ambient vibration tests were performed to obtain the dynamic characteristics of tall reinforced concrete shear wall buildings with the height range from 15 to 45 stories, located in downtown Vancouver, Canada. The raft foundation's vibration levels, mode shapes, modal frequency, damping, rocking behavior, and soil–structure interaction were determined (Turek et al. 2008). It was observed that the fundamental

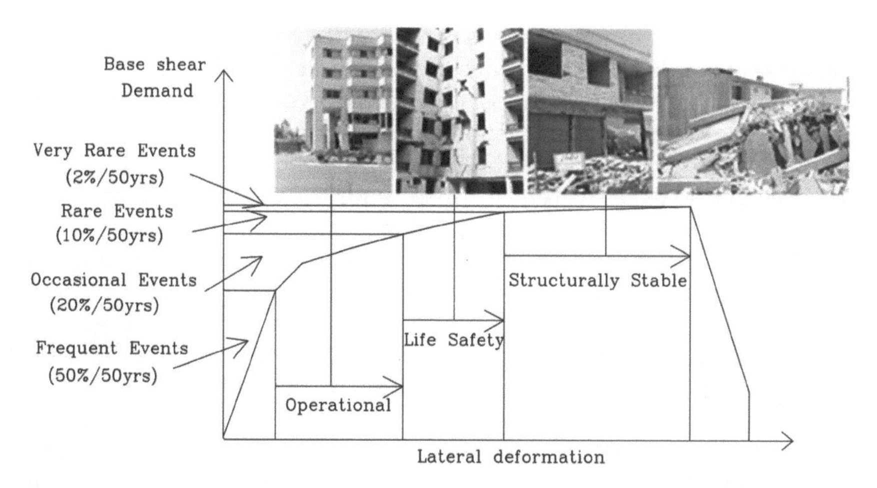

Figure 6-1. A schematic view of selecting a performance level.
Source: Adapted from Hamburger and Holmes (1998). Photo: F. M. Tehrani, buildings in Rasht and Manjil after the 1990 Manjil-Rudbar Earthquake (Tehrani 1990).

natural period increased from 0.81 to 3.57 s. These results showed that NBCC 2005 is more conservative for taller buildings, and by using the predicted period from finite element modeling (FEM), lower design forces can be used for response spectrum analysis. Regarding the micro-tremor and base analysis results, the movement of the foundation and the performance of the structure's vibration modes had a significant correlation. It showed a high capacity for soil–structure interaction effects because of the rocking of the foundation.

The 8.8 magnitude Maule earthquake in 2010 caused strong ground motion, where most of the affected buildings were reinforced concrete (RC) structures. It was observed that most structural damages were in multistory and high-rise buildings because of the poor performance of slender RC shear walls without confined boundary elements that caused crushing of concrete and buckling of vertical wall reinforcement at the end and throughout the entire length of the wall. Because Chilean code did not provide any restrictions in designing irregular structures, soft and weak stories and discontinued shear walls existed, resulting in increased force and deformation demand with global and local failures. Moreover, the interaction of nonstructural components with the seismic force-resisting system caused damage. For instance, unexpected forces applied to surrounding columns around window openings provided by masonry walls created short columns, resulting in diagonal tension failures. Some failures were due to a lack of proper connections in precast structural elements and tilt-up walls (Saatcioglu et al. 2012).

Four ductile steel moment-resisting 5-, 10-, 15-, and 20-story frame buildings were designed in Vancouver, Canada, and their performance was investigated to assess the seismic safety level specified in the code. The buildings were initially designed using equivalent static load for the seismic load according to NBCC 2005 (NRCC 2005), and then the modal and dynamic analyses were performed. Equivalent seismic lateral forces in inverted triangular form as provided in NBCC 2005 (NRCC 2005) were used for the pushover analysis, and these forces were also monotonically increased for the analysis. The nonlinear dynamic response of these structures was assessed using synthesized and scaled actual ground motion records. The results of pushover and dynamic analyses indicated that building frames designed according to the NBCC 2005 seismic provisions achieved the anticipated performance level of collapse prevention or better. The pushover analysis revealed that an increase in the building height reduces the ductility capacity of a building frame's system. The ductility capacities of the 15- and 20-story buildings were much lower than those of the 5- and 10-story frames, which were more than five, as was assumed to determine the design lateral forces. Infill panels decreased the dynamic drift demand and damage, as well as the structural system ductility capacity. In addition, building performance was affected by the nature of selected ground motion records (Yousuf and Bagchi 2009).

6.1.6 Control Systems

Three significant control systems are available: passive control system, active control system, and semiactive control system. Combining these control systems is the so-called hybrid control systems consisting of combined passive and active devices or passive and semiactive devices.

6.2 ACTIVE CONTROL SYSTEMS

Active control systems require a significant power source because electrohydraulic or electromechanical actuators generate the control forces based on feedback information from the measured response of the structure or external excitation. These measurements are monitored by a controller, which determines the control signal for the operation of the actuators.

6.3 PASSIVE ENERGY DISSIPATION DEVICES

Passive control systems do not require an external power source for operation and are used to modify the dynamic properties of a structure, thus reducing the demand on the structural system. Supplemental energy dissipation devices can take many different forms and use different

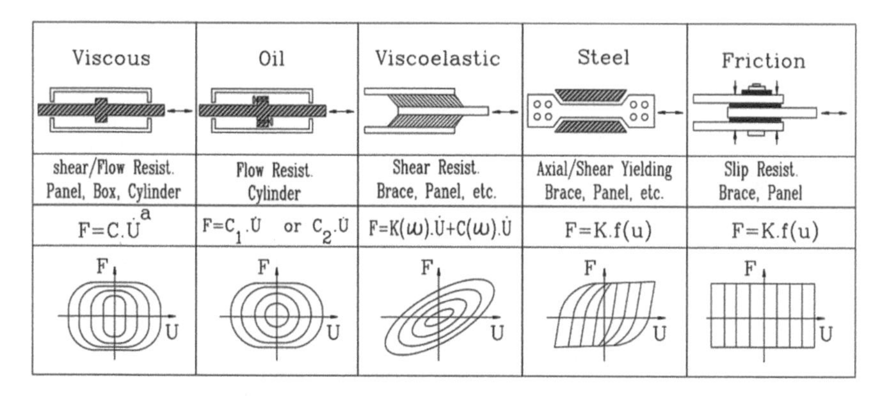

Figure 6-2. Significant types of damper technology.
Source: Adapted from Kasai et al. (2007).

mechanisms to dissipate energy, including the yielding of mild steel, viscoelastic action in rubber-like materials, shearing of viscous fluid, orificing of fluid, and sliding friction (Symans et al. 1997). The seismic isolation system is one of the most well-accepted passive control devices. A flexible isolation system is placed between the foundation and the superstructure to increase the natural period of the system; this results in reducing acceleration in the superstructure, increasing the displacement in the isolation level. When incorporated into a structure, passive energy dissipation devices reduce energy dissipation demands on primary structural members and minimize possible structural damage by absorbing a portion of the input energy.

Passive energy dissipation devices can be divided into six groups: metallic dampers, friction dampers, viscoelastic dampers, viscous fluid dampers, tuned mass dampers, and tuned liquid dampers (Constantinou et al. 1998). Figure 6-2 shows major types of dampers. The primary function of viscous and oil dampers is to resist the flow of polymer liquid and low viscosity oil. The hysteresis of a viscous damper is a combination of ellipse and rectangle and can be modeled in series combination with nonlinear dashpot and elastic spring, whereas the series combination of linear dashpot and elastic spring is used to model the oil damper. The inclined elliptical shape is developed using a viscoelastic damper, which dissipates energy using a polymer's molecular motion. Energy in steel and friction dampers is dissipated by the yielding of steel material and through the friction between two solid bodies sliding off next to each other (Kasai et al. 2007).

6.4 SEMIACTIVE CONTROL SYSTEMS

A small external power source is required for operating semiactive control systems, and a controller monitors the feedback, generates an

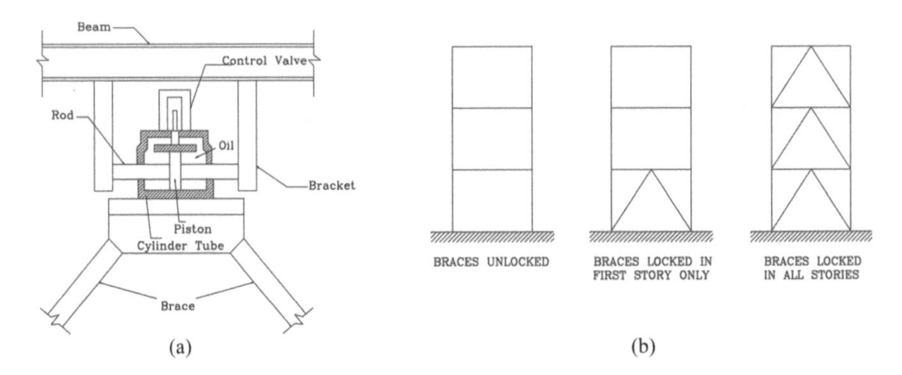

Figure 6-3. Stiffness control device (a) installation detail, (b) configurations within a full-scale test structure.
Source: Adapted from Kobori et al. (1993).

appropriate command signal for the devices, and utilizes the structural motion to create the control forces. Stiffness control devices, electrorheological dampers, magnetorheological (MR) dampers, friction control devices, fluid viscous dampers, tuned mass dampers, and tuned liquid dampers are some examples of these systems.

The primary function of stiffness control devices is to modify the stiffness and, thus, natural vibration characteristics. Figure 6-3 shows the semiactive stiffness device in a chevron bracing arrangement. The origins of this device can be traced to a hydraulic cylinder with a solenoid control valve. When the valve is closed, the beam locks to the braces below, and when it is open, it disengages the beam and the brace connections; at each time step, the stiffness configuration is specified, and suitable command signals are sent to the stiffness control devices. Analytical results of this system indicate that during lower-magnitude earthquakes (e.g., 4.9 on the Richter scale), there was a 70% reduction in the roof acceleration of the structure; this amount decreased to 40% for higher-magnitude events (e.g., 5.7 on the Richter scale).

Electrorheological (ER) dampers consist of a hydraulic cylinder containing micron-sized dielectric particles suspended within a fluid. On the existence of a strong electric field, the particles polarize, become aligned, and produce an increased resistance to flow. The behavior of ER dampers can be modulated by changing the electric field. A large-scale capacity ER was developed (McMahon and Makris 1997), as demonstrated in Figure 6-4. It was established that when the case has no applied electric field, the elliptical shape of the hysteresis loop can be modeled as a linear viscous dashpot.

MRs are the magnetic analogs of ER dampers with similar behavior but with a magnetic control effect instead of an electric control effect. MR dampers consist of a hydraulic cylinder containing micron-sized,

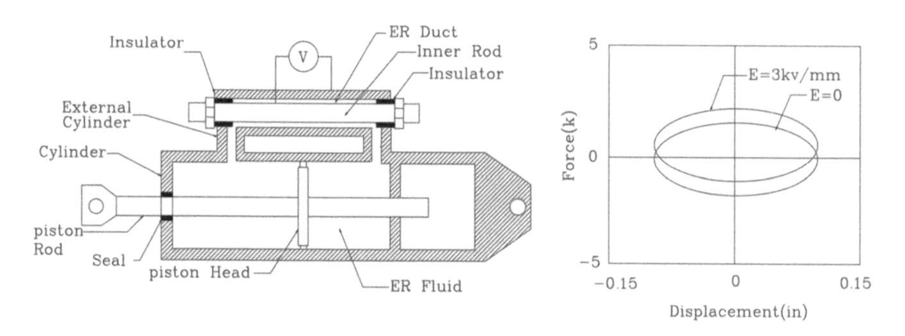

Figure 6-4. Schematic of a small-scale damper and hysteresis loop for a large-scale damper for two different electric field strengths.
Source: Adapted from McMahon and Makris (1997).

magnetically polarizable particles suspended within a fluid. The system's performance is tested by subjecting the fluid to a magnetic field.

Semiactive friction control devices are used as energy dissipaters within the lateral bracing or as components within sliding isolation systems. The idealized hysteresis loop of the friction damper is presented in Figure 6-5. As the force increases, the hysteresis loop expands vertically; thus, the normal force controls the amount of dissipated energy per cycle of harmonic motion.

An isolation system was described by Feng et al. (1993) to limit the sliding displacement and minimize the transfer of seismic forces to the superstructure, in which the friction force on the sliding interface between the superstructure and the foundation is controlled. Figure 6-6 demonstrates a cross-sectional and plan view of the semiactive friction control bearing.

As shown in this figure, each bearing has a fluid chamber and a pressure control system, which consists of a servo valve, an accumulator, and a computer used to modify the pressure.

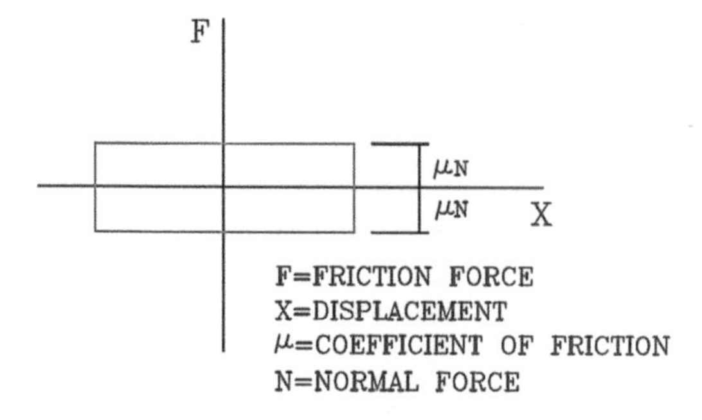

Figure 6-5. Idealized Coulomb-friction-damper hysteresis loop.
Source: Adapted from Feng et al. (1993).

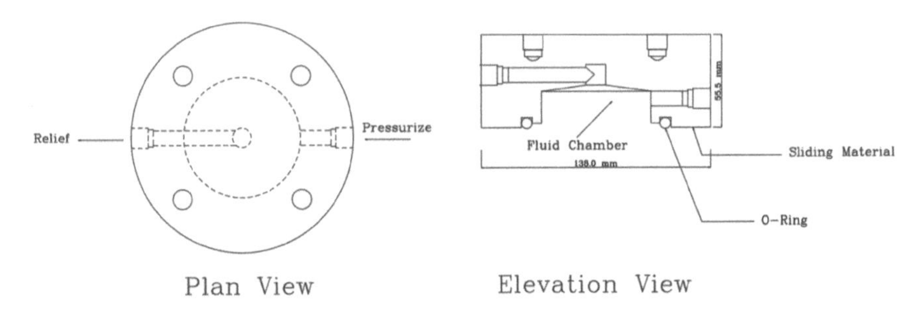

Figure 6-6. Friction-controllable bearing.
Source: Adapted from Feng et al. (1993).

Semifluid viscous dampers consist of a hydraulic cylinder with a piston head to separate the two sides of the cylinder. When the piston is cycled, the fluid is compelled to pass through small orifices at high speed and the pressure variation across the piston head, and an external control valve modulates the output force; this control valve is in the form of a solenoid valve for on–off control or a servo valve for variable control. Symans et al. (1997) tested two different semiactive damper systems: two-stage and variable dampers utilizing solenoid and servo valves, respectively. The semiactive fluid damper includes a stainless-steel piston rod, a bronze piston head, and a piston rod makeup accumulator and is filled with thin silicone oil, as shown in Figure 6-7.

A single degree of freedom (mass–spring–damper) can be used to model tuned mass dampers. A tuned mass damper is mounted on the top floor of a multistory structure, and the motion of the structure is controlled by tuning the dynamic characteristics of the system. Tuned liquid dampers are like tuned mass dampers, except that a container filled with fluid replaces the mass–spring–damper system. Lou et al. (1994) recommended

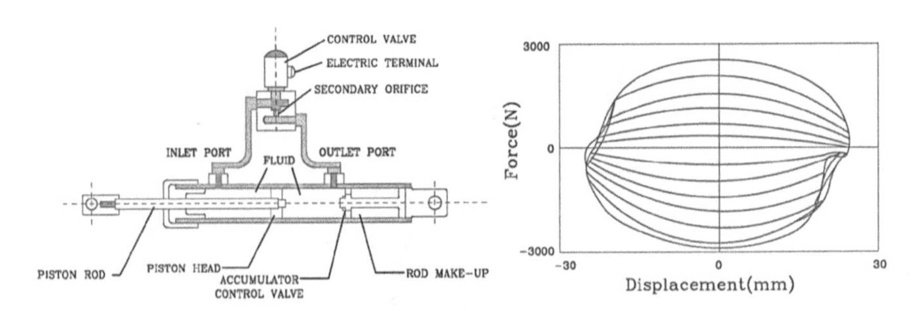

Figure 6-7. Schematic view of damper and hysteresis loops for seven different command voltage levels subjected to harmonic motion.
Source: Adapted from Symans and Constantinou (1997).

a semiactive tuned liquid damper. The behavior of the semiactive damper and the natural frequency of the sloshing fluid are controlled by the length of a hydraulic tank and by adjusting the position of rotatable baffles in the tank, respectively; this indicates the benefits of different tank lengths for controlling the response of the mass.

It was observed from the experimental testing that the semi-active control method could improve the seismic behavior of structures. With a comparison of structural control systems (i.e., passive, active, and semiactive), it was noted that semiactive control systems are a better option because they do not have the limitations of passive and active dampers. However, the large-scale semiactive control systems for seismic response reduction need further investigation.

6.5 RESILIENCE OF ELASTOMERIC- AND FRICTION-BASED SYSTEMS

Elastomeric- and friction-based devices are passive energy dissipation systems, as noted in prior sections. The significant advantages of these systems include providing flexibility at the base of the structure, reducing the story drift, increasing the natural period of the structure, and consequently, reducing the design base shear of the structure, as schematically presented in Figure 6-8 (Tehrani 1996). Enhancement of the flexibility of the structure is the critical contribution of isolating devices to

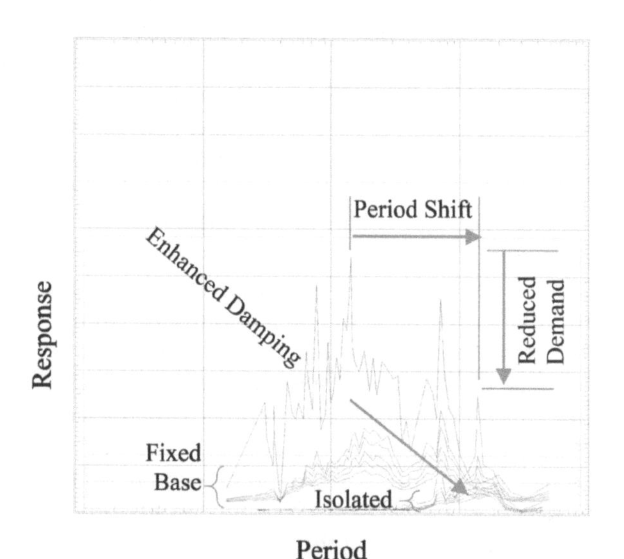

Figure 6-8. Concept and mechanisms of seismic isolation.

reduce transferred forces from the foundation to the structure. Reductions of story drift, natural frequency, and design load are consequences of the isolation mechanism (Ferritto 1991, Mayes et al. 1988, Tehrani 1992, 1993a, Tehrani and Hassani 1994a, Golafshani et al. 1996).

Standard components of the isolation mechanism include three essential elements: first, an isolating mechanism to provide flexibility; second, a damper to control the relative displacement of the building with respect to the ground; and third, a resisting system to withstand minor lateral loads when the service of isolation is not necessary. A fourth component (i.e., a self-centering mechanism) is common in specific isolation systems (Tehrani 1993c). The first component, i.e., a flexible or isolating element, typically shifts the structure's natural period to higher levels, where the earthquake demands subside. The lower demand will reduce story drifts as well. However, such an increase in the structure period will increase the total displacement of the structure, which may result in stability issues. Thus, the presence of the second element (i.e., a damper) is necessary to control such displacements by dissipating the energy of the earthquake. The design of the third element, i.e., a resisting system to withstand slight wind and earthquake events, is proportional to the capacity of the structure for such loads. The rigidity and strength of this system may prevent the structure from returning to the original location after a significant event, which may be an operational concern. In this case, a self-centering mechanism will be beneficial (Buckle 1989, Buckle and Mayes 1990, Mayes et al. 1988, Tehrani and Hassani 1994b, 1995, Hassani et al. 1996). Rubber bearings are the most basic elastomeric-based systems with applications in many buildings and bridges for vibration control. In this category, natural rubber bearings offer an isotropic behavior with equal stiffness in all directions, which is not the best fit for isolating structures from horizontal ground motions (Hassani et al. 2001). The next significant advancement, the layered rubber bearing with horizontal reinforcing steel plates, provides an opportunity to flex the bearing horizontally without a substantial reduction of vertical stiffness. Furthermore, the addition of a lead core to this reinforced bearing enhances the damping ratio and is instrumental in controlling the deformation of the bearing. The lead-rubber system satisfies the need to withstand minor lateral loads without drift, isolate the structure at the initiation of a strong ground motion, and dampen the energy of the earthquake throughout the event. Figure 6-9 shows the schematic section of rubber bearings with the lead core (Tehrani 1992, 1993c).

Figure 6-10 presents a simplified two-dimensional model of an isolated system with elastomeric bearings (Tehrani 1993b, Tehrani et al. 1993, 1994). In this model, the masses of the roof and base floor are comparable. The deformed shape is presented in solid lines as opposed to dashed lines for the undeformed shape. It can be shown that the relative

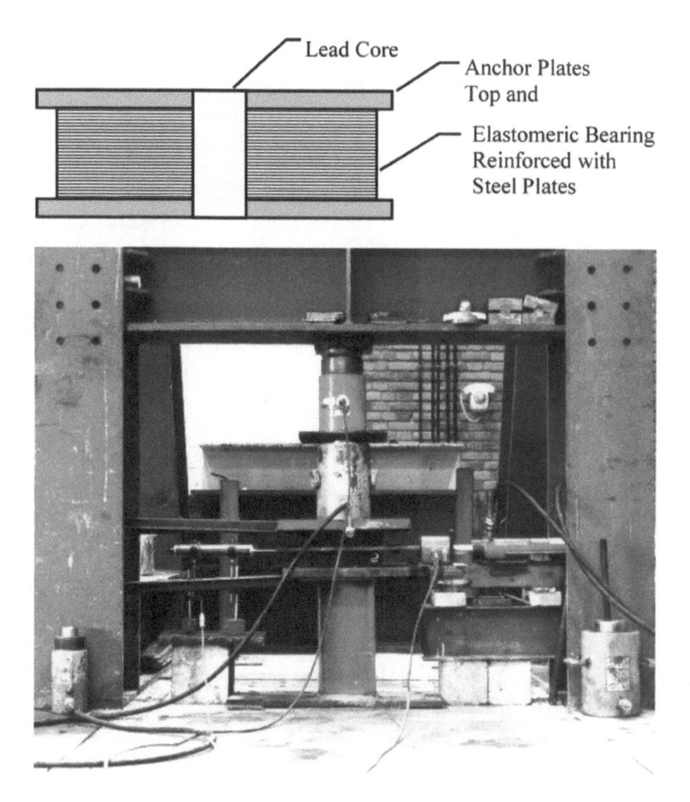

Figure 6-9. Typical elastomeric bearing system with (top) lead core and (right) testing.
Source: F. M. Tehrani, elastomeric testing using a strong frame photo (Hassani et al. 2001).

displacement of the top of the structure in respect to the base, also known as the story drift, is much less than the relative displacement of the base with respect to ground movement if the stiffness of the structure is substantially lower than the stiffness of the bearing. Further, implementing large damping values at the base concerning negligible damping of the structure will contribute to energy dissipation at the base (Kelly 1999). It is also evident that the isolation has shifted the natural period of the system from 0.14 s for the fixed base structure to 1.5 s for the base-isolated structure (Tehrani 1993b).

Sliding isolators or pure-friction isolation systems represent another class of isolators (Tehrani 1993b, Tehrani and Ahmadi 2015). The friction mechanism can simply act as the isolator when the transferred force from the ground to the structure exceeds the friction resistance in the system. Then, the friction drives the damping mechanism by dissipating energy while the superstructure slides on top of the substructure. The static

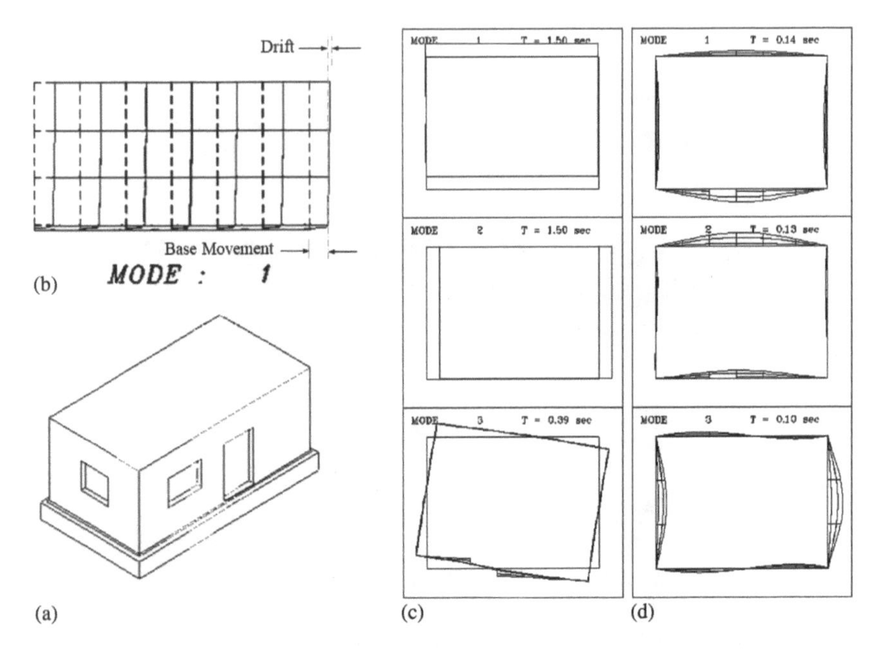

Figure 6-10. (a) 2D model of an isolated structure, (b) drift and base movement of an isolated structure, (c) modal analyses for isolated, (d) fixed base models.

friction of a sliding isolator also locks the system during minor events, which are not expected to trigger the system. Figure 6-11 presents three approaches to measuring the coefficient of friction as the essential parameter in the behavior of pure-friction isolation systems (Tehrani 1993b, Hassani et al. 1996).

The simple mechanism of pure-friction isolation systems has prompted their application in low-cost and low-tech structures such as adobe buildings (Tehrani 1993b, Tehrani and Massoud 1995, Tehranizadeh and Tehrani 1997, Hassani et al. 1996). Figures 6-12 and 6-13 present two examples of sliding layers using different granular materials and system configurations (Li 1984, Tehrani 1993b, Tehrani and Hassani 1996). Figure 6-14 presents a constructed building, the first in Iran, using a pure-friction isolation system (Tehrani 2000).

A common challenge in the pure-friction system is the lack of a self-centering system to restore the structure to its original location after a seismic event. The friction pendulum system addresses this problem by providing a spherical friction surface that allows an articulated slider to move against a polytetrafluoroethylene-bearing material (Figure 6-15) (Naeim and Kelly 1999, Zayas et al. 1989).

Advantages of using the friction mechanism to dampen the energy of ground motions have resulted in hybrid systems of elastomeric and

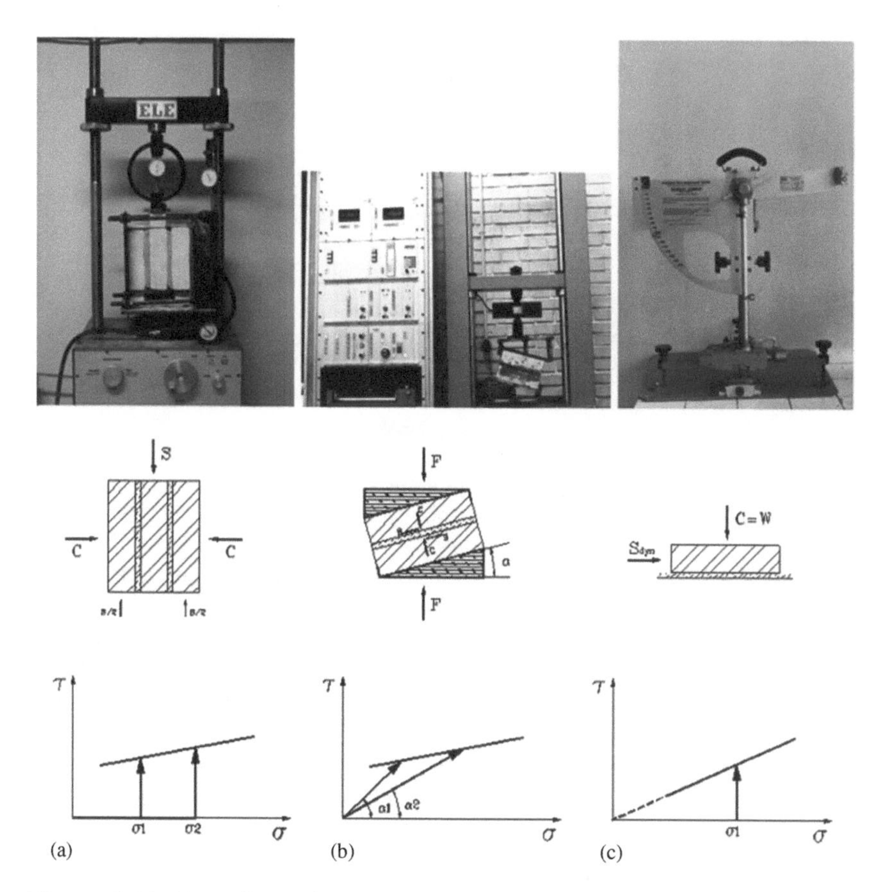

Figure 6-11. Experimental measures to determine the coefficient of friction: (a) biaxial fixed-lateral pneumatic load method at a 1.27–50.8 mm/min rate, (b) fixed-angle wedge method at a 50.8 mm/min rate, (c) dynamic fixed normal load method at an 80 m/min rate.
Source: (a) Tehrani (1992), (b) Hassani et al. (1996), (c) Tehrani (1993b).

friction-based devices. The resilient-friction base isolation (R-FBI) system (Mostaghel 1984) is an example of such hybrid systems, where Teflon-coated stainless-steel flat rings are assembled around a central rubber core (Figure 6-16). The friction between these plates enhances the damping characteristics of the device. Figure 6-17 shows another hybrid device developed by Electricité-de-France, where the sliding occurs at the top of the main assembly of laminated neoprene bearings because of using a top stainless-steel plate. Because of the separation of friction and elastomeric-based mechanisms in the EDF system, the sliding of the superstructure on top of the stainless-steel plate can act as a fuse, similar to pure-friction systems that isolate the structure from severe ground motions. This system

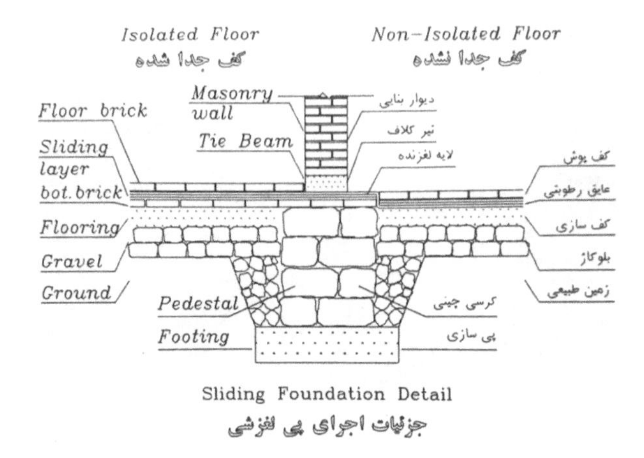

Figure 6-12. Application of a sand layer between flat plates.
Source: Adapted from Tehrani (1993b, 1998).

has been applied to protect essential structures, such as nuclear power plants, built near active faults (Jangid and Kelly 2001).

A combination of the R-FBI and EDF systems has also been considered to enhance damping. An example of such a system is the sliding resilient-friction (S-RF) system. This system uses the R-FBI device as the base of the EDF system and, therefore, takes advantage of friction on the top, as the fuse, and between plates, as the damper (Su et al. 1991). Another advanced method of enhancing damping is applying hysteretic damper devices with high damping ratios known as the NZ (New Zealand) system.

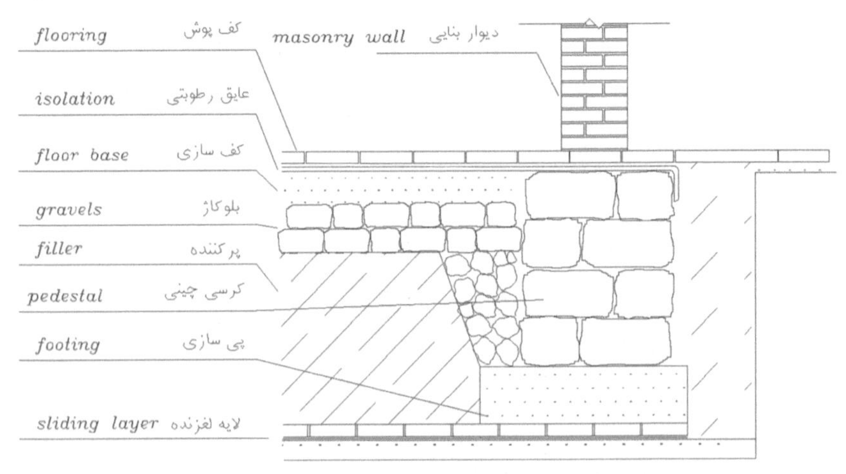

Figure 6-13. Application of lightweight-expanded clay aggregates under foundation.
Source: Adapted from Tehrani (1993b, 1998).

Figure 6-14. The first pure-friction base-isolated masonry building in Iran. Photo: F. M. Tehrani, LECA Plant, Saveh, Iran.
Source: Adapted from Tehrani (2000).

Figure 6-18 compares schematic diagrams for selected types of isolation systems supporting a rigid mass. These diagrams indicate how friction, elastomeric, and damping mechanisms are assembled to perform as an isolator unit. These diagrams have been expanded to represent the behavior of shear beams and multistory buildings, in addition to the simplified rigid mass model, as shown in Figure 6-19 (Su et al. 1989, 1990, Fan et al. 1990a, b, 1991).

The response of various structural systems to selected ground motions indicates that isolation systems are, in general, effective in reducing transferred forces and story drifts (Figures 6-20 and 6-21). Therefore, seismic isolators can enhance the resilience of the system by reducing structural damage during an earthquake event. Further, the gap between

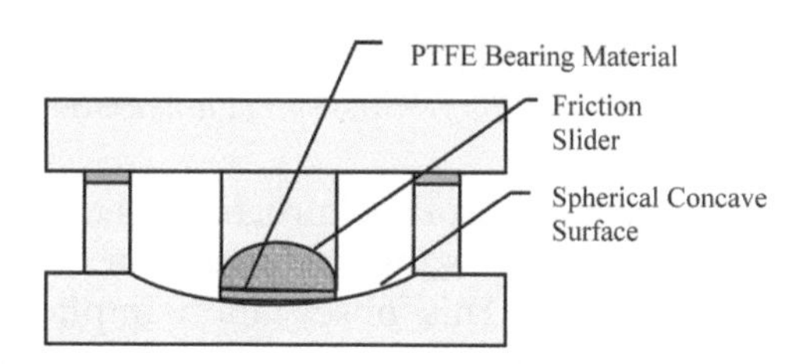

Figure 6-15. Schematic of a friction pendulum system.

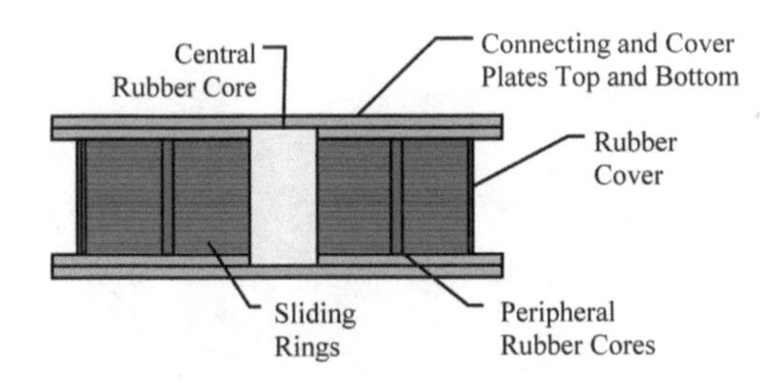

Figure 6-16. Schematic of an R-FBI system.

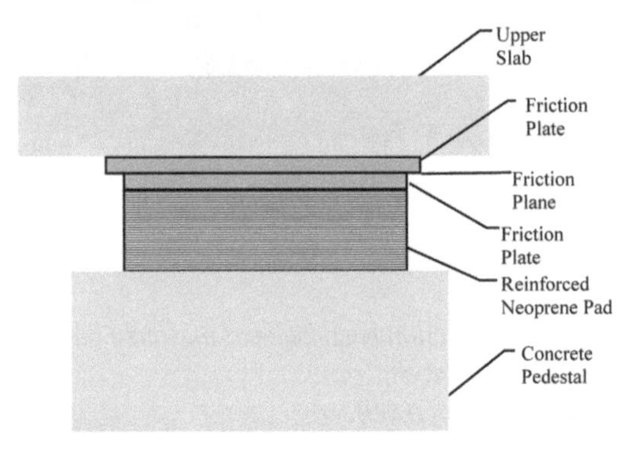

Figure 6-17. Schematic of an EDF system.

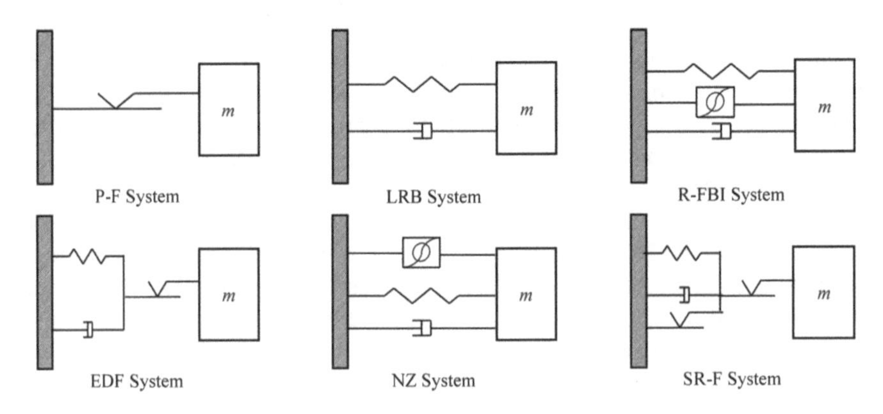

Figure 6-18. Schematic diagrams of selected isolation systems.
Source: Adapted from Su et al. (1989, 1990), Fan et al. (1990a, b, 1991).

the responses of fixed-base and isolated structures is more significant for stronger ground motions, such as the Pacoima Dam event, with higher ground acceleration values. This observation implies that isolated buildings are more resilient to unforeseen ground motions with severe intensities. However, base-isolated structures are sensitive to the

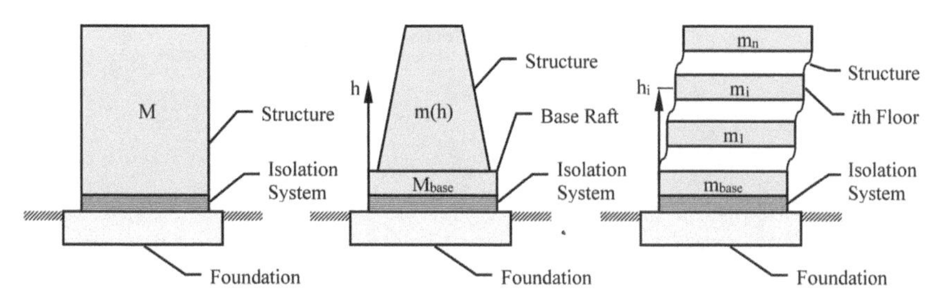

Figure 6-19. Schematic diagrams of selected isolation systems.
Source: Adapted from Su et al. (1989, 1990), Fan et al. (1990a, b, 1991).

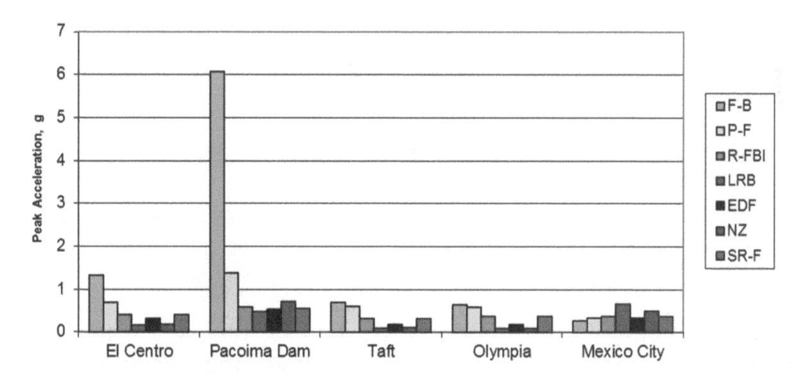

Figure 6-20. Sensitivity of seismic isolation systems in respect to the peak acceleration to selected ground motions.
Source: Adapted from Su et al. (1990).

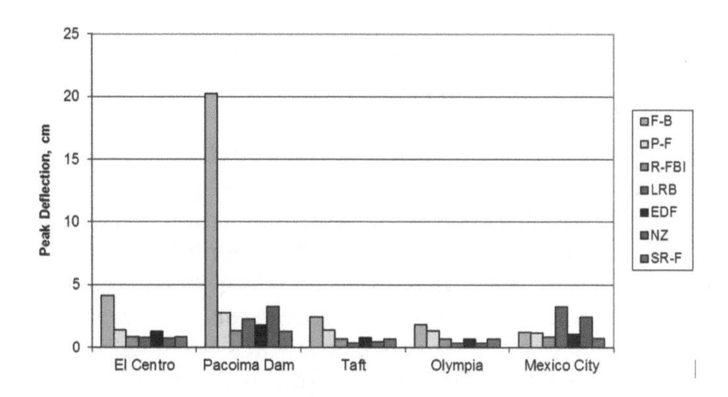

Figure 6-21. Sensitivity of seismic isolation systems in respect to the peak deflection to selected ground motions.
Source: Adapted from Su et al. (1990).

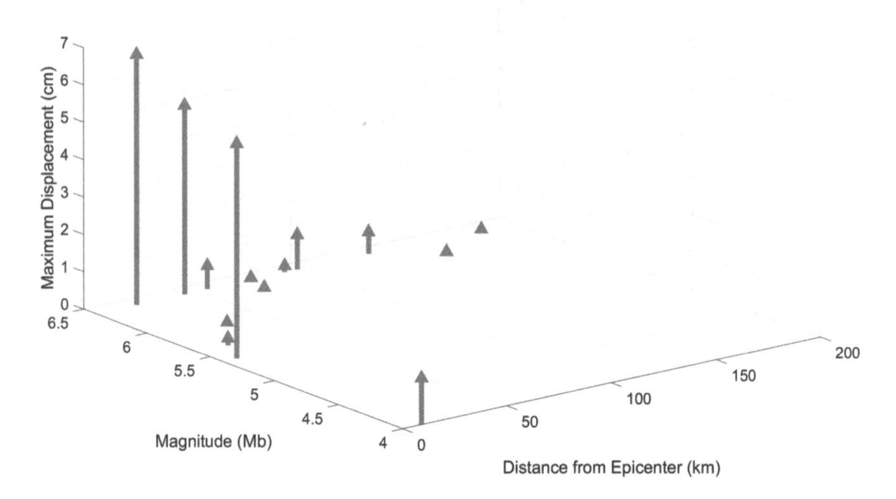

Figure 6-22. Sensitivity of pure-friction isolation systems in respect to the base displacement for selected ground motions.

frequency content of the ground motion. This sensitivity was the case for the Mexico City event, where a long period of ground excitation resulted in a larger response of isolated structures than that of the fixed-base structure. This sensitivity is a significant concern for elastomeric-based devices when the frequency content of the earthquake is close to the vibration period of the device. The addition of friction systems, in general, alleviates such consequences by showing less sensitivity to the ground motion characteristics such as frequency content. Therefore, hybrid systems, in general, perform better when subjected to a portfolio of different ground motions and contribute to the resiliency of the system (Tehrani 1992).

Tehrani (1998) extended studies on the sensitivity of pure-friction base isolation systems to the magnitude and the distance from the epicenter of selected ground motions. Figures 6-22 and 6-23 indicate that pure-friction isolation systems with a low coefficient of friction of 0.1 (in comparison with the typical values of 0.2) may experience large base movements and/ or large accelerations because of the vertical component of the ground motion during major seismic events possibly at a magnitude of above 6.0, within a 50 km distance to the epicenter, or during minor seismic events, possibly at a magnitude between 4.0 and 6.0, within a 10 km distance to the epicenter (Tehrani 1998).

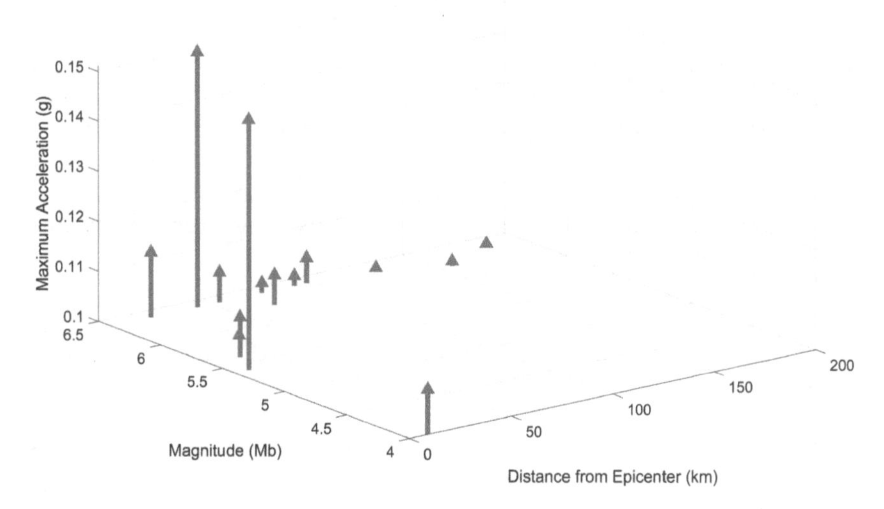

Figure 6-23. Sensitivity of pure-friction isolation systems in respect to the acceleration for selected ground motions.

6.6 RESILIENCE OF LOW-DAMAGE SELF-CENTERING SYSTEMS

In recent years, self-centering or rocking systems have been widely used in structural systems. Although this concept has been used in different types of structures, it was primarily used in concrete systems to reduce the significant damage concentrated at plastic hinges. A joint US–Japan research program titled PREcast Seismic Structural Systems (PRESSS) was initiated in the early 1990s (Cheok and Lew 1991, Priestley and Tao 1993, Priestley et al. 1999) toward constructing seismic-resilient options for precast structures. In this program, self-centering connections were developed for improving the seismic performance of beam–column joint subassembly and wall-to-foundation connections in precast concrete buildings. By using post-tensioning (PT) tendons as the primary reinforcement of beam-to-column and wall-to-foundation connections, the lateral resistance of the elements is provided and their recentering capability are secured as the PT tendons are designed to remain elastic within the design drift level of the structure. In the cast-in-place (CIP) structures, these connections were initially designed to nonlinearly respond by allowing the formation of plastic hinges in the region; this provides a large amount of energy dissipation, as hinges act as fuses for the structure, as well as consequent residual deformation, crack formation, and therefore severe damage in the system. In self-centering systems, sufficient energy dissipation capacity, which is the crucial component of a resilient structure, is lacking because of the omission of the plastic hinge formation. Therefore, several external energy dissipating elements are integrated into the system to work with the self-centering elements. In what follows, first, the self-centering systems, their lateral load resistance, and energy dissipation mechanisms are discussed. Then, a literature review on the topic is presented, which addresses the historical perspective, up-to-date research findings, and codification. Finally, different types of these systems and their applications in existing structures around the world are discussed.

6.7 SELF-CENTERING SYSTEMS: LATERAL LOAD RESISTANCE AND DAMPING MECHANISMS

The application of self-centering systems in all types of structures has been practiced, but these elements were mainly used in concrete precast structures to eliminate major concrete cracking as the material responds nonlinearly. In what follows, lateral load resisting and energy dissipation components of self-centering frames and wall systems in concrete structures are discussed. Then, the applications of self-centering systems in steel and timber self-centering structures are presented.

As shown in Figure 6-24 (a, b), self-centering beam-to-column or wall-to-foundation connections are constructed by joining the corresponding structural elements using unbonded (or partially bonded) PT tendons. The tendons, which are aimed to respond linearly with no residual deformation up to the design drift of the system, elongate when the system is subjected to any lateral loading. Because of this, the structure experiences the opening of a single crack at the self-centering connections, resulting in a rocking response for the elements at the rocking interface. The prestressing force mainly provides the lateral load resistance of these systems. It is separated from their energy dissipation mechanism, which, in general, is maintained through natural inherent viscous damping of the system and negligible impact damping owing to the rocking of the elements at the self-centering connection. This mechanism results in a resilient performance of the system in terms of generating negligible residual drift; however, the system encompasses slight damping capacity while it allows for input energy during external lateral loading (e.g., earthquake). Several types of energy dissipating elements were added to these systems to overcome the lack of energy dissipation capacity in self-centering concrete structures and provide the required damping and maintain the system's resiliency. Examples are U-shaped flexural plates (UFP) (Priestley et al. 1999), grouted mild steel reinforcement (Rahman and Restrepo 2000, Holden 2001), and

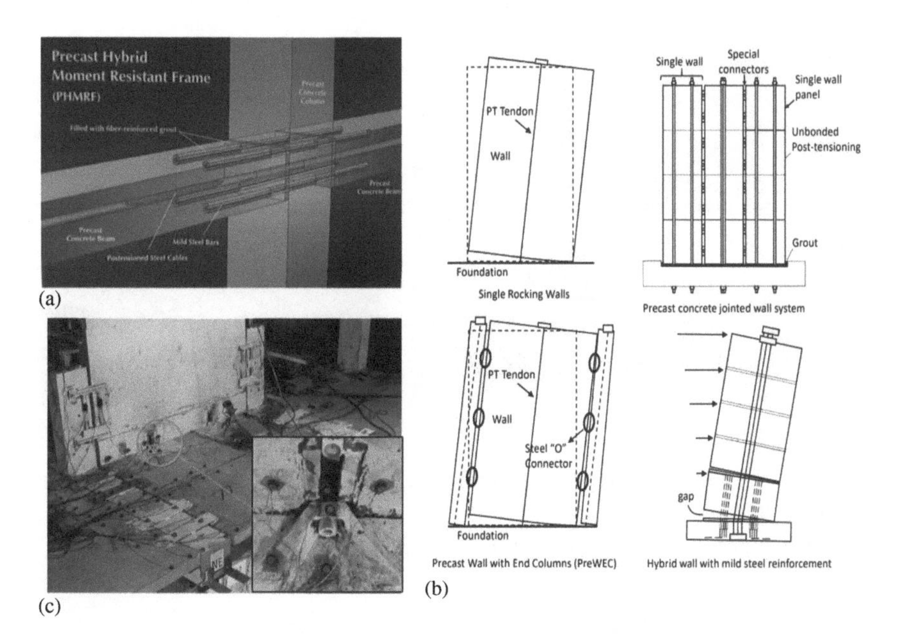

Figure 6-24. Self-centering connections in resilient concrete structures: (a) Self-centering beam-to-column connections, (b) self-centering rocking wall systems with and without energy dissipating elements, (c) resilient connection of the rocking wall system with the surrounding structure.
Source: (a) Englekirk (2002), (b) Kurama et al. (1999), Priestley et al. (1999), Rahman and Restrepo (2000), Sritharan et al. (2008), (c) Liu (2016).

oval-shaped mild steel connectors (Sritharan et al. 2008, 2015). As shown in Figure 6-24(b), these steel connectors are used to join the self-centering wall panels together or to the surrounding structural elements (e.g., foundation or columns); these connectors are designed to yield, due to rocking of the self-centering wall panel and provide additional hysteretic damping for the system. These elements, which are, in most cases, easily replaceable after the lateral excitations, also provide external lateral load resistance for the rocking system. Interactions of the self-centering rocking wall systems with the surrounding structure have always been a challenging topic. A few studies (e.g., Fleischman et al. 2005, Henry 2011, Liu 2016) suggested applying special connectors to provide a resilient wall-to-floor connection. An example of such connection (Liu 2016) is shown in Figure 6-24(c).

Figure 6-25 indicates the application of different self-centering systems in steel and timber structures. As presented in this figure, similar concepts adopted for concrete structures are used in these systems to post-tension the elements at the rocking joints (Pampanin 2015, Lin et al. 2009).

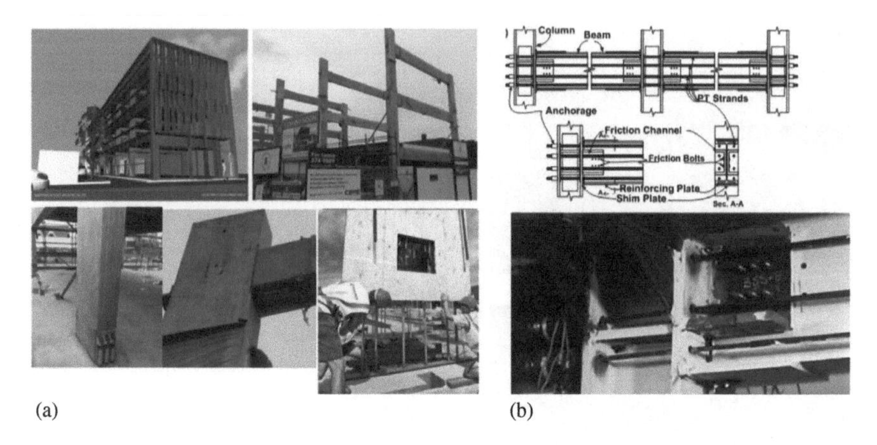

Figure 6-25. Self-centering connections in a resilient: (a) timber structure, (b) steel structure.
Source: (a) Pampanin (2015), (b) Lin et al. (2009).

6.8 HISTORICAL PERSPECTIVE

Housner (1963) initially modeled the rocking behavior as he attempted to illuminate the reasons behind the survival of many tall, slender structures in the 1960 Chilean earthquake (Henry 2011). Modern structural engineers have investigated this concept over the last 28 years. In 1991, the application of fully grouted post-tensioned beam-to-column connections in the initial tests that were carried out as part of the inter-program coordination of the US PRESSS program by the National Institute of Standards and Technology confirmed its inappropriate performance in terms of the provided damping capacity (Cheok and Lew 1991). Following this practice, Priestley and Tao (1993) suggested applying unbonded PT tendons to join precast members of structural frames and adding special spiral reinforcements at the end regions of beams to modify the limited damping capacity of the self-centering connections. In addition, during the PRESSS program, researchers at Lehigh University analytically investigated the concept of using unbonded PT connections in rocking walls (Kurama et al. 1999). These walls were later equipped with additional hysteretic energy dissipaters to be used in high seismic regions. Examples are jointed or coupled walls (Priestley et al. 1999), hybrid walls (Rahman and Restrepo 2000, Holden 2001), and precast walls with end columns (PreWECs) (Sritharan et al. 2008, 2015), as previously shown in Figure 6-24(b). Resilient connection of these wall systems to the floor slabs was also practiced using slotted insert shear connectors during shake table testing of a prototype precast concrete parking structure at the University of California at San Diego (UCSD) (Fleischman et al. 2005). Recent research on this topic is discussed in the following section.

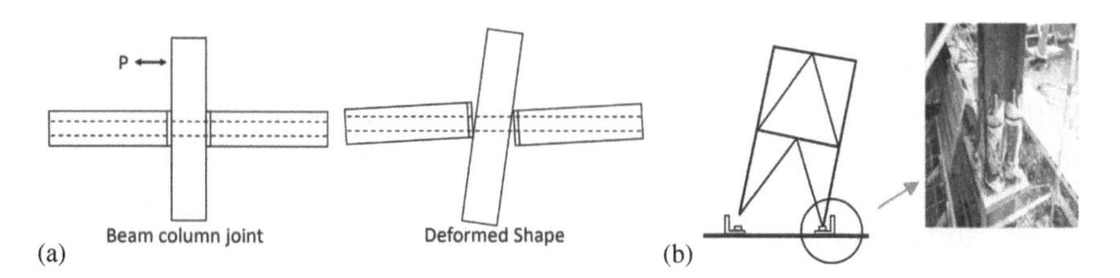

Figure 6-26. Self-centering steel structures: (a) post-tensioned beams in steel moment frames, (b) rocking steel concentrically braced frame.
Source: Adapted from (a) MacRae (2010), (b) Chanchí et al. (2010), Sidwell (2010).

Application of post-tensioned beams in steel moment frames and rocking steel concentrically braced frames, as shown in Figure 6-26, was studied by different researchers (e.g., Danner and Clifton 1995, Clifton 2005, Christopoulos et al. 2002, Roke et al. 2006). The self-centering concept was also used for timber structures, mainly by the researchers in New Zealand (e.g., Palermo et al. 2005, Buchanan et al. 2008) to limit the structural damage by assuring the self-centering capability of the system.

6.9 RECENT RESEARCH AND CODIFICATION

ACI ITG-5.1 (ACI 2007) summarizes the seismic acceptance criteria for special unbonded post-tensioned precast concrete walls based on previous studies on concrete wall systems, including large-scale, quasi-static testing and analytical studies accomplished as of 1999 (e.g., Priestley et al. 1999, Kurama et al. 1999, Stanton and Nakaki 2002). In addition, relying on experimental and analytical research, ACI ITG-5.2 (ACI 2009) presents specific design provisions for jointed and hybrid unbonded post-tensioned wall systems. These codes have recently been incorporated into ACI 550.6 and 550.7 (ACI 2019a, b). In 2006, Appendix B of the New Zealand concrete design standard, i.e., NZS 3101-06 (NZS 2006), outlined appropriate provisions for designing unbonded PT precast concrete elements.

Although the application of self-centering systems can result in less residual deformation and damage of the buildings, the seismic application of these systems (especially without the application of external dampers) has been limited because of the uncertainty about their energy dissipation capacity. Various researchers (e.g., Marriot 2009, Nazari 2016, Nazari et al. 2017, Twigden 2016, Nazari and Sritharan 2018, 2019, 2020) have conducted shake-table studies to evaluate the performance-based response of these self-centering systems subjected to different intensities of ground motions. Nazari et al. (2017) and Nazari and Sritharan (2018, 2019, 2020) showed that different types of rocking walls with and without external energy–dissipating elements might be designed following the current design codes, for example, ASCE 7-16 (ASCE 2017), with recommended response modification coefficients (or R factor) corresponding to their energy dissipation ratio and respond satisfactorily in terms of maximum lateral drift, maximum absolute acceleration, and residual drift during design basis earthquake (DBE) and maximum considered earthquake (MCE) events. The uncertainty in the interaction between the self-centering system and the surrounding structure has also limited the implementation of these self-centering systems. However, recent studies (e.g., Liu 2016) provided different types of wall-to-floor connections to minimize the local damage that occurred to slabs because of rocking of the self-centering elements during seismic action. Although not implemented in the design codes yet, these recent research studies shed light on the dynamic behavior of self-centering concrete buildings in high seismic regions.

6.10 WORLDWIDE APPLICATIONS

Self-centering systems have been used in the construction of several buildings located in high seismic regions worldwide. Those built in New Zealand were even tested under real earthquake shaking during the 2011 Christchurch event. The 39-story Paramount building in downtown San Francisco, California (Englekirk 2002) is constructed using self-centering hybrid moment–resisting frames. The Southern Cross Hospital, located in Christchurch, New Zealand, is one of many buildings that has been constructed using self-centering systems, and it is noteworthy that this building resisted the 2011 Christchurch earthquake. This building used the post-tensioned hybrid walls coupled with U-shape flexural plate dissipaters, and as shown in Figure 6-27, there was minimal damage to both structural and nonstructural components of the building (Pampanin et al. 2011).

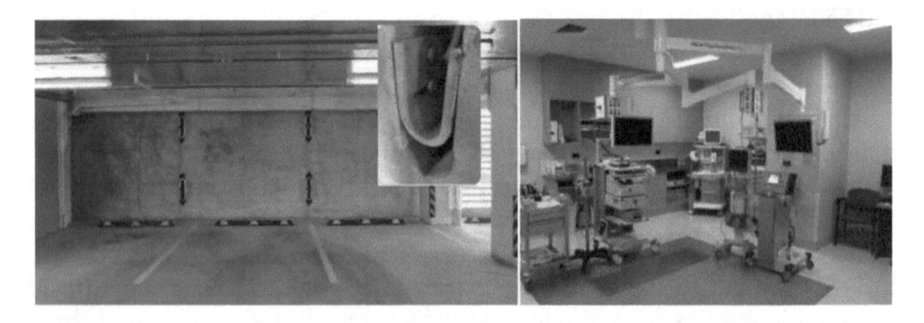

Figure 6-27. Negligible observed damage to the Southern Cross Hospital designed with a rocking-dissipative solution during the Christchurch earthquake. Source: Pampanin et al. (2011).

6.11 RESILIENCE OF ROCKING FOUNDATIONS

Rocking systems were commonly used in the construction of superstructures in buildings and bridges. For bridge systems, several studies have been done to incorporate rocking foundations to reduce the damage in the substructures (e.g., Espinoza and Mahin 2006, Kawashima and Nagai 2006). The ability of soil under the rocking foundation to dissipate energy through soil yielding was also studied through analytical and experimental investigations by several researchers (e.g., Harden and Hutchinson 2009, Negro et al. 1998).

Through numerical investigations on six three-dimensional reinforced concrete bridges in Oakland, California, Antonellis and Panagiotou (2014) demonstrated elastic response and small residual deformations of the bridge models with rocking foundations. Figure 6-28(a) demonstrates a schematic of their model. Saad et al. (2012) conducted a shake table study on a two-fifth scale of a three-span highly curved bridge. They mentioned that the effect of the rocking foundation on the seismic responses is similar to the case where the seismic isolation is used for the system. As discussed in Kawashima and Nagai (2006), this is because of the significant reduction of the plastic deformation of the column at the plastic hinge zone [see Figure 6-28(b) for their modeling approach]. They also advised to properly select the footing size, as the seismic isolation effect can enlarge the displacement demands of the superstructure. Allmond and Kutter (2014) studied the application of rocking foundations in poor soil conditions and were notified that this may lead to a significant settlement accumulation. To overcome this issue, they presented the application of unattached piles for this type of soil condition and proper design considerations for such systems.

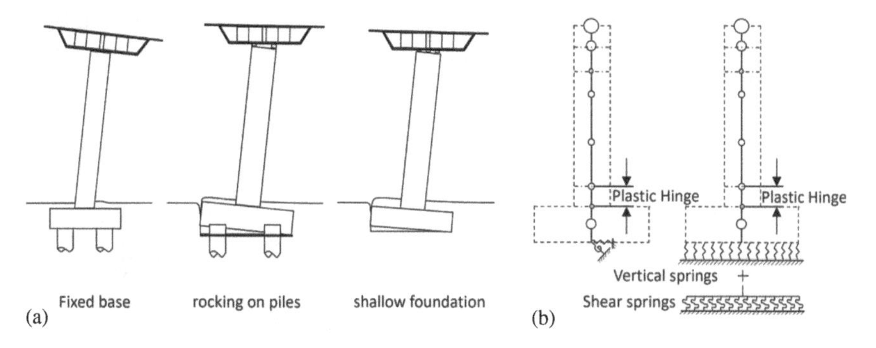

Figure 6-28. Rocking foundation systems: (a) Schematic of foundation, column, and deck in a deformed state, (b) idealization of spread foundation without and with separation of the footing from the underlying ground.
Source: Adapted from (a) Antonellis and Panagiotou (2014), (b) Kawashima and Nagai (2006).

6.12 RESILIENCE OF ADVANCED FRICTION DAMPERS

Friction damper dissipates energy through the friction between two solid bodies sliding off next to each other. Earthquake generation and tectonic movement can be controlled by solid friction as an instance. Automotive brake is another example on a smaller scale that can dissipate the kinetic energy of motion. The friction brake is commonly used to withdraw kinetic energy from a moving body (Pall and Pall 1996). When a major earthquake occurs, the friction damper slips before yielding at a fixed load in the members of a frame, which dissipates a substantial amount of energy. The friction damper saves the initial cost of new constructions or retrofitting existing buildings, with very high energy dissipation.

Several devices have been developed to dissipate energy through friction, including limited slip bolt joint (Pall et al. 1979), X-braced friction damper (Pall and Marsh 1982), Sumitomo friction damper (Aiken and Kelly 1990), energy dissipating restraint (Aiken et al. 1992, Nims et al. 1993a), and slotted bolted connection (FitzGerald et al. 1989). The following literature provides a brief explanation of some friction devices.

6.12.1 Limited Slip Bolt Joint

In large panel structures, the damage is usually located along the joints during an earthquake; therefore, the joints are practically the only locations where energy dissipation occurs. Based on the concept of energy dissipation, Pall (1979) maximized their capacity and developed a dissipated joint for seismic control of large panel structures, shown in Figure 6-29.

The limited–slip bolted (LSB) joint design incorporated brake lining pads between steel plates to provide a consistent force–displacement response. Pall (1979) conducted several experimental tests under static and dynamic

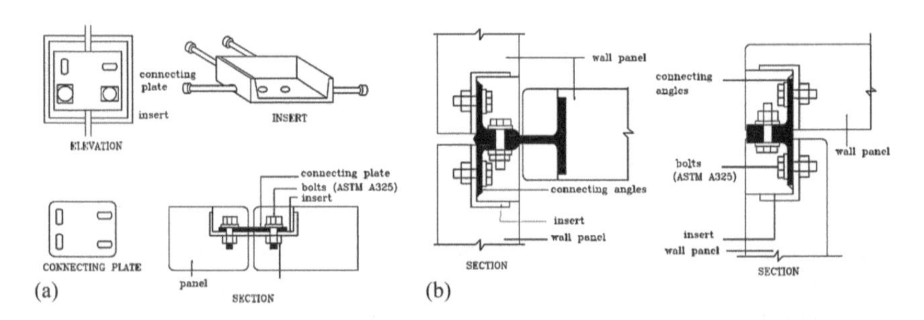

Figure 6-29. The LSB joint: (a) wall-to-wall joint, (b) corner wall-to-wall joint. Source: Adapted from Pall (1979).

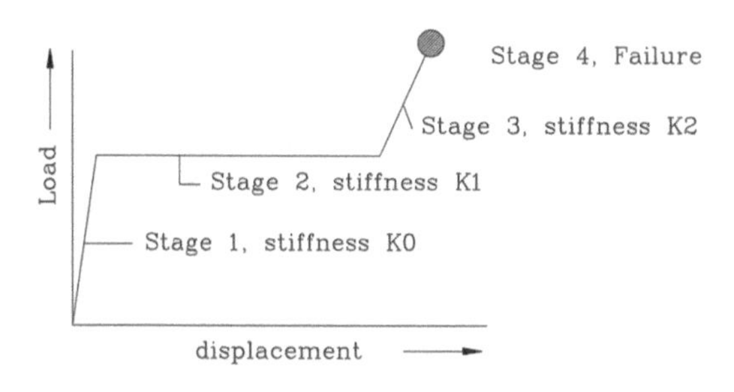

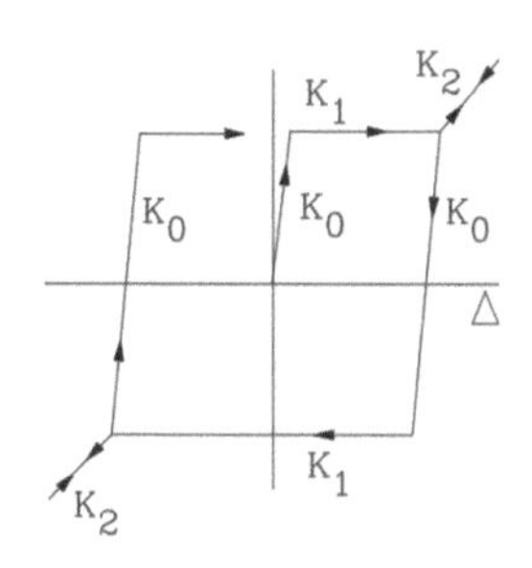

Figure 6-30. Idealized behavior of the LSB Joint.
Source: Adapted from Pall et al. (1980).

cyclic tests to obtain primary design data and realistic structural response on a difference of simple sliding components having various surface treatments, including mill-scale, sand-blasted, inorganic zinc-rich paint metalized brake lining pads, and a polyethylene coating. Although metalized surfaces showed the highest static slip coefficient and energy dissipation, their performance was far from predictable. The best behavior was shown by brake lining pads located between steel plates with a mill-scale surface.

Based on the behavior obtained by Pall et al. (1979), the idealized behavior of the LSB joint is shown in Figure 6-30, where the slipping phase is shown in Stage 2 and simulated by the plateau.

6.12.2 Three-Stage Friction Grip Elements

Roik et al. (1988) discussed seismic control of structures under earthquake loading by three-stage friction-grip elements. The energy dissipation of each story can be designed based on serviceability and medium- and strong-motion earthquakes based on the friction joints, which Pall (1984) verified. The idea was created from braking by friction; for instance, the driver can softly push the pedal to avoid high forces by sudden braking.

Tests performed on a single joint were investigated for simple steel–concrete- and steel–steel-friction-grip connections to show the mechanism of energy dissipation (SFB A51), as shown in Figure 6-31. The durability of this type of connection under high short-term dynamic loading is highly satisfactory.

Friction grip joints can resist high dynamic loading using nonlinear spring elements with bolts controlling the loss of prestressing, and their hysteresis depends on the prestressing of the bolts, as shown in Figure 6-32.

Roik et al. observed that coupling in parallel, as shown in Figure 6-33(a), prevents vibration by the transition phase from elastic behavior to the slipping stage. The component behavior and the predicted performance of a three-stage stiffening element, as well as hysteresis loop, are presented

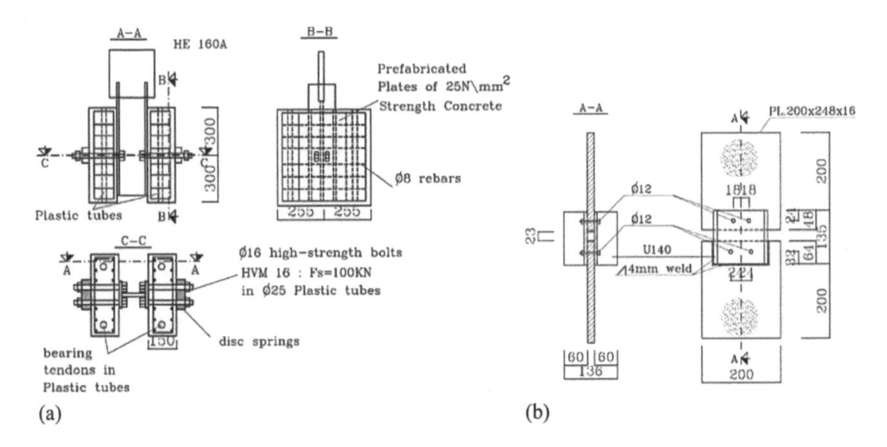

Figure 6-31. (a) Test specimen for steel–concrete, (b) steel–steel-friction-grip connections.
Source: Adapted from Roik et al. (1988).

in Figure 6-33(b–d). Stage 1 is the serviceability limit state for linear structural behavior and small displacement. Stage 2 is the transition stage with no damage and larger displacement under a medium-intensity earthquake to obtain the required smooth transition from Stage 1 to Stage 3. Stage 3 is the ultimate design limit state under the maximum strong motion with minor damage and large displacement.

A seven-story building was chosen for the three-system investigation: ductile frame, three-stage truss, and stiff core, as shown in Figure 6-34. The truss system with three-stage elements was modeled by using the girders hinged to columns. All three models were calculated by using 5% damping and the P-delta effect.

As shown in Figure 6-35, the elastic concrete core has a small maximum displacement, and the three-stage truss limits both the horizontal

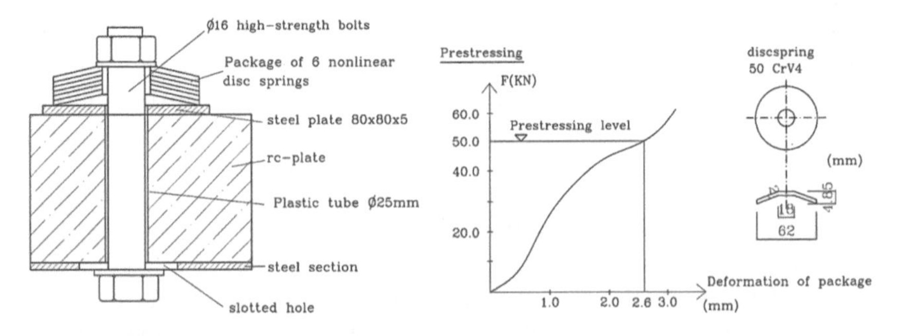

Figure 6-32. Detail of steel-concrete friction grip by a nonlinear disk spring.
Source: Adapted from Roik et al. (1988).

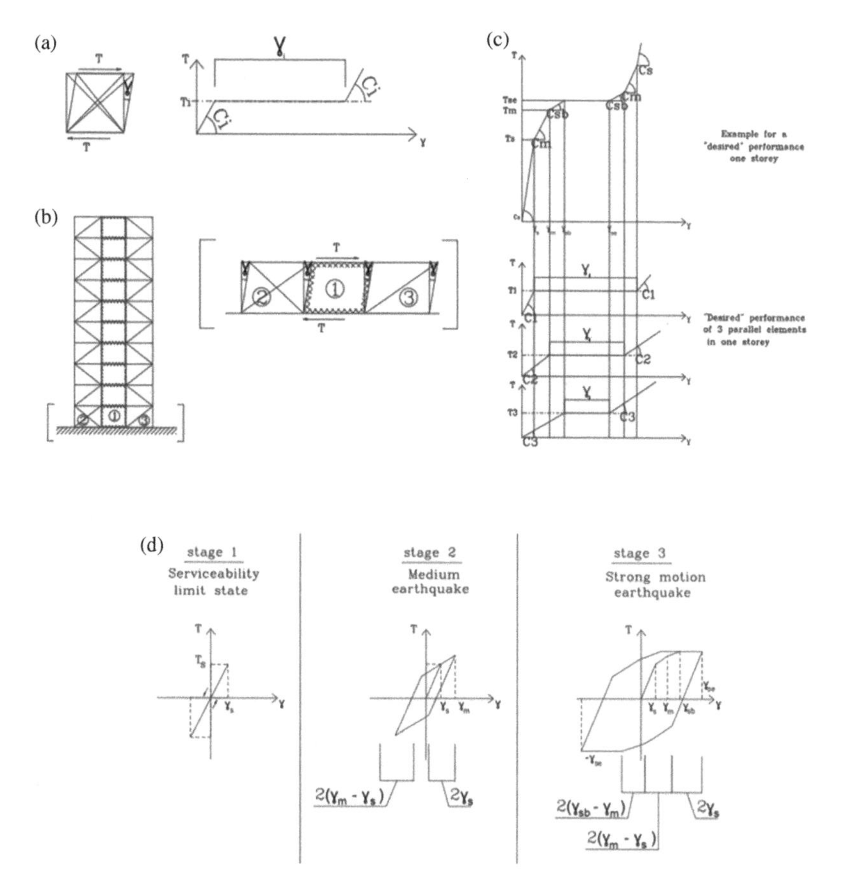

Figure 6-33. (a) Principal behavior of one stiffening element: Ci: stiffness; γ_i: frictional displacement; Ti: frictional force (level of friction), (b) three displacement coupled parallel stiffening elements, (c) three-stage stiffening element, (d) hysteresis loops of a three-stage stiffening element.
Source: Adapted from Roik et al. (1988).

displacement and the story shear because of energy balance versus time. Because the internal forces of both the three-stage truss and the ductile frame are similar, they show the same energy balance, with the frame having a higher kinetic and viscous energy. The three-stage truss shows the highest percentage of energy dissipation, which allows an economical design.

A simulated earthquake helped ascertain the behavior of the three-stage truss, and the specimen was scaled down with the available testing facility, as shown in Figure 6-36.

The mechanical properties of bolts and the limited geometry of slotted holes play an important role in lateral stiffness, slip force, and the amount

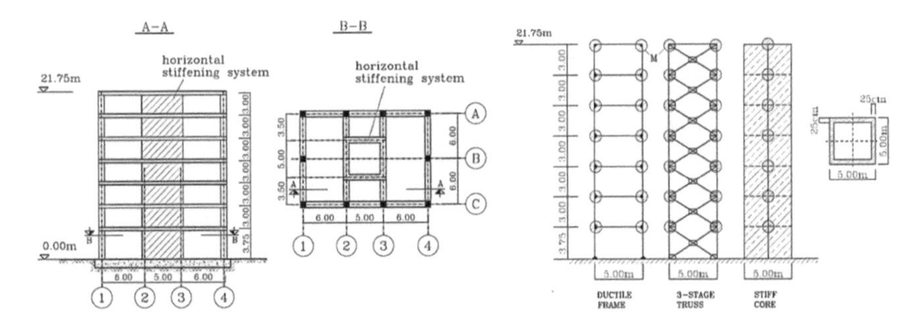

Figure 6-34. Seven-story building and the three investigated versions.
Source: Adapted from Roik et al. (1988).

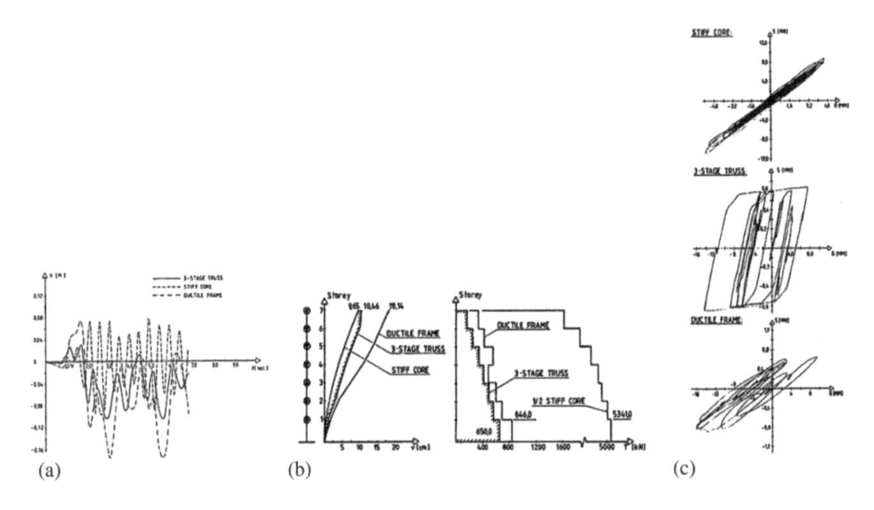

Figure 6-35. (a) Roof displacement versus time, (b) maximum horizontal displacement and shear force, (c) hysteresis loop for the first story of the three systems.
Source: Adapted from Roik et al. (1988).

of dissipated energy per cycle. It was concluded that friction grip connections could reduce horizontal forces and displacements by their energy dissipation capacity. The experimental test was limited to one story, and the effect of the whole structure was not considered; moreover, three-stage elements using steel concrete friction grip required further study.

6.12.3 Friction Dampers

Based on the development of LSB, Pall and Marsh (1982) recommended a system in which the braces in a moment-resisting frame incorporated frictional devices. During severe earthquake excitations, a large portion of

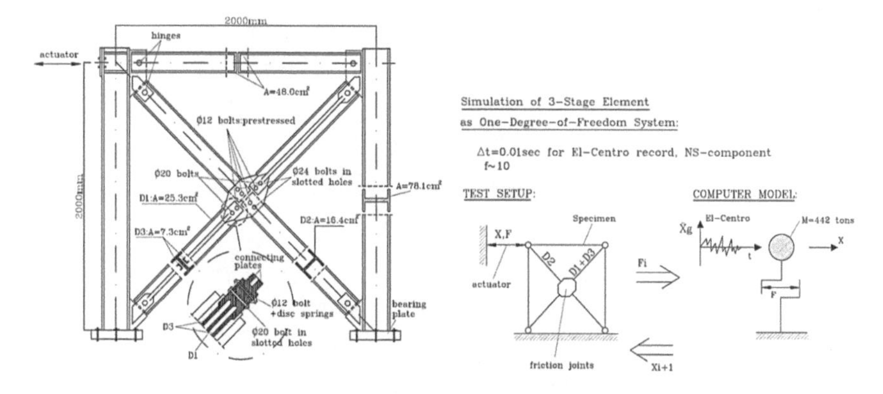

Figure 6-36. Specimen setup for a three-stage truss and the simulation of one degree of freedom.
Source: Adapted from Roik et al. (1988).

the energy is dissipated by the friction when the device slips. Friction dampers can be grouped as follows (Pall and Pall 1996): friction damper in X-bracing, friction damper in chevron bracing, and friction damper in single diagonal, as shown in Figure 6-37 (Couch 2020).

Friction dampers have large rectangular hysteresis loops, behaving like an ideal elastoplastic mechanism, and their performance is independent of velocity. The forces on the members can be controlled to be within their capacity by using an appropriate slip load. Pall friction dampers in line and cross bracing were used to upgrade the Boeing commercial airplane factory; these dampers notably reduced the forces exerted and lateral deflections. Therefore, the strengthening of the existing members was minimized, and this plan provided significant savings in terms of both construction cost and time (Vail et al. 2004). A chevron brace with two friction dampers was used for the Sharp Memorial Hospital. The results of the nonlinear analysis showed an economic performance-based design.

Friction dampers have been utilized as a practical and cost-effective energy dissipation mechanism in many constructed structures. Chang et al. (2006) examined the application of friction dampers for the seismic retrofit of a three-story steel structure that did not satisfy the current building code seismic requirement. Because the third floor functioned as court premises, 48 friction dampers were used on the ground and second floor.

FEMA 351 and 356 were used in the analysis using a three-dimensional (3D) model by ETABS, and the story shear and friction dampers reduced displacement as demonstrated in Figure 6-38.

It was observed that the friction dampers significantly improved the structural performance, and with this reduction in story displacement and shear, the seismic force was reduced above the third floor.

Figure 6-37. Braced frame using a Quaketek friction damper.
Source: F. M. Tehrani, an experimental shake table study photo (Couch 2020).

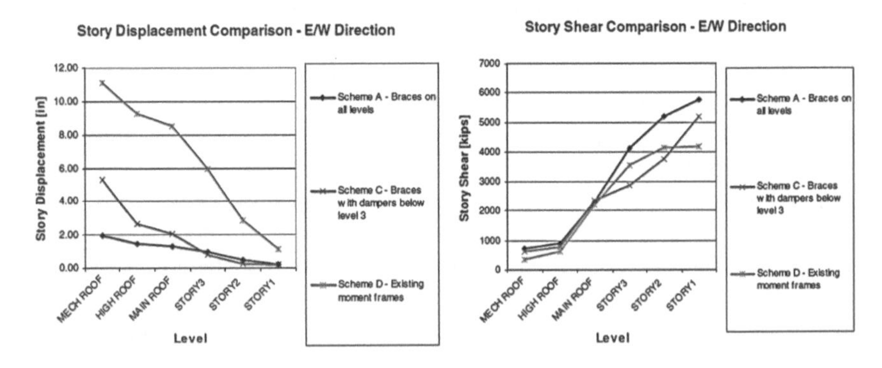

Figure 6-38. Story displacement and shear comparison.
Source: Adapted from Chang et al. (2006).

Patil et al. (2015) examined a performance-based plastic design for a 21-story steel moment–resisting frame with friction dampers. Nonlinear static pushover analysis and nonlinear time history analysis were performed to achieve a uniform target drift agreeing with peak inter–story drift limits. It was observed that using friction dampers can help specify a certain amount of inelastic displacement for a given earthquake; moreover, there was a reduction of almost 85% in the peak inter–story ratio and the flexural moment in the columns.

Colajanni and Papia (1997) examined the hysteresis characteristic of one-story friction damped brace frames to evaluate the role of the period of vibration, the lateral stiffness ratio, and the global slip load calibration of the dissipative device. It was observed that the frequency of the slip excursions relies on the period of the system vibration and the average amplitude of the normalized slip excursions is independent of this period and increases in terms of the lateral stiffness ratio. The improvement in the response of friction-damped multistory frames required further study regarding the distribution of the overall slip load of the devices and the structure's height.

Morgan et al. (2007) investigated utilizing friction dampers in the seismic design of unbounded post-tensioned precast concrete frame structures. A nonlinear-reversed cyclic analysis of post-tensioning steel areas and selected damper distribution was performed to evaluate the slip forces of the friction dampers subjected to lateral load. It was shown that friction-damped precast concrete frames could maintain a high level of self-centering capability because of the post-tensioning force while dissipating energy levels.

A tall cylindrical tower, which is used in industrial applications, is called a process column. However, during a seismic assessment, it was observed that anchor bolts were not meeting the code requirement. Therefore, a retrofitting scheme using passive control devices was used (Kiran et al. 2016). Different passive control devices, including a viscoelastic damper, elastoplastic damper, tuned mass damper, and tuned liquid damper and friction, can be easily replaced after a ground motion. Because viscoelastic dampers are affected by temperature and stiffness degradation, tuned dampers, which have some limitations, are required for tuning the natural frequency. Instead of these tuned dampers, double-sliding friction dampers were used as they do not possess the limitations of tuned dampers. Figure 6-39 shows a typical friction damper and its end connections.

The cyclic load was repeated for 10 cycles by considering various torque values, a stable hysteretic behavior was obtained, and the differential of slip load was linear concerning the applied torque. After retrofitting, the seismic demand was reduced to 15% of the capacity of the existing foundation bolts, and the structure was qualified to satisfy the MCE condition.

The damping mechanism combination of a nonlinear Reid damper and a viscous damper showed that the response of the structure could be reduced by the piezoelectric friction damper (PFD) (Chen and Chen 2001). Numerical simulations also showed that friction dampers driven by what

Figure 6-39. Friction damper.
Source: F. M. Tehrani, an experimental shake table study photo (Couch 2020).

is suggested as control logic could considerably reduce the peak acceleration and story drift of a building structure subject to ground motion excitation. The same results were obtained by Haider and Kim (2012). The effects of friction dampers subjected to several seismic excitations were studied, and it was revealed that the friction dampers effectively dissipate a significant amount of energy. A semiactive electromagnetic friction damper was proposed by Agrawal and Yang (2000). The recommended damper and the control method were studied for a base-isolated building, which showed that the semi-active electromagnetic friction damper (SAEMFD) was functional in preserving the rubber bearings of the base-isolated buildings under strong earthquake excitations.

6.13 RECOMMENDATIONS

1. Plan: Seismic isolation and protective devices significantly contribute to the resilience of structures. The state of practice and research in this area provides a broad range of systems for new and existing structural systems and nonstructural components. The selection of appropriate systems typically depends on buildings' architectural and structural features, site and hazard characteristics, target performance and service time span, and available technologies. The resultant geostructural interaction during a seismic event determines the overall contribution of the isolation system and protective devices to the resilience of the infrastructure.

2. Implementation: The main component of a seismic isolation system is the elastomeric or sliding isolator that reduces or eliminates the transferred load to the structure. This component practically contributes to the reduction of vulnerability and the enhancement of safety. Further, selecting a system with less sensitivity to hazard characteristics, like intensity and frequency, improves the reliability of the system's performance in response to future uncertain events. Enhanced reliability contributes to the robustness and resilience of the structure.
3. Implementation: The damping mechanism in isolation and protective devices reduces damage through the dissipation of the ground motion energy. Sacrificing devices are popular because of their ease of repair and replacement, which also enhances the recovery and rapidity of the capacity restoration, subject to the availability of the implemented technology. Attention to resourcefulness is essential for advanced and active systems that rely on external sources for materials or energy.
4. Implementation: Isolated structures often require a built-in or added mechanism to provide strength against nonextreme loadings. These mechanisms may also contribute to the system redundancy against aftershocks. However, over-specification of isolation systems may reduce the efficiency of such systems.
5. Assessment: The selected target in the performance-based design of isolated and protected structures provides a measure for assessing damage in such structures. Such assessment applies to both structural and nonstructural components. Implemented self-restoring and active mechanisms also benefit from such assessments throughout the service time span of the infrastructure. Further, hybrid systems enhance the resiliency of systems exposed to a series of different seismic events.

BIBLIOGRAPHY

Calugaru, V. 2013. "Earthquake resilient tall reinforced concrete buildings at near-fault sites using base isolation and rocking core walls." Ph.D. thesis, University of California, College of Civil Engineering.

Calvi, G. M., T. J. Sullivan, and D. P. Welch. 2014. "A seismic performance classification framework to provide increased seismic resilience." In *Perspectives on European earthquake engineering and seismology*, 361–400. Cham, Switzerland: Springer.

Chen, M. C. 2018. "Enhanced seismic resiliency for buildings via base isolation." Ph.D. thesis, University of California, Dept. of Structural Engineering.

Chiaro, G., A. Palermo, G. Granello, A. Tasalloti, C. Stratford, and L. Banasiak. 2019. "Eco-rubber seismic-isolation foundation systems: A cost-effective way to build resilience." In *Proc., Auckland: 11th Pacific Conf. on Earthquake Engineering and 2019 NZSEE Conf.*, Paper 196: 1–8. Auckland, New Zealand: Australian Earthquake Engineering Society.

Dall'Asta, A., G. Leoni, F. Micozzi, L. Gioiella, and L. Ragni. 2020. "A resilience and robustness oriented design of base-isolated structures: The new Camerino University Research Center." *Front. Built Environ.* 6: 50.

De Domenico, D., and G. Ricciardi. 2018. "Earthquake-resilient design of base isolated buildings with TMD at basement: Application to a case study." *Soil Dyn. Earthq. Eng.* 113: 503–521.

Dong, Y., and D. M. Frangopol. 2016. "Performance-based seismic assessment of conventional and base-isolated steel buildings including environmental impact and resilience." *Earthq. Eng. Struct. Dyn.* 45 (5): 739–756.

Du, X. L., Y. L. Zhou, Q. Han, and Z. L. Jia. 2019. "Shaking table tests of a single-span freestanding rocking bridge for seismic resilience and isolation." *Adv. Struct. Eng.* 22 (15).

Etedali, S., K. Hasankhoie, and M. R. Sohrabi. 2020. "Seismic responses and energy dissipation of pure-friction and resilient-friction base-isolated structures: A parametric study." *J. Build. Eng.* 29: 101194.

Farsangi, E. N., A. A. Tasnimi, T. Y. Yang, I. Takewaki, and M. Mohammadhasani. 2018. "Seismic performance of a resilient low-damage base isolation system under combined vertical and horizontal excitations." *Smart Struct. Syst.* 22: 383–397.

Forcellini, D. 2017. "Seismic resilience of isolated bridge configurations with soil–structure interaction." *Innov. Infrastruct. Sol.* 2 (1): 2.

Giouvanidis, A. I., and Y. Dong. 2020. "Seismic loss and resilience assessment of single-column rocking bridges." *Bull. Earthq. Eng.* 18 (9): 4481–4513.

Guo, T., Z. Xu, L. Song, L. Wang, and Z. Zhang. 2017. "Seismic resilience upgrade of RC frame building using self-centering concrete walls with distributed friction devices." *J. Struct. Eng.* 143 (12): 04017160.

Hashimoto, T., K. Fujita, M. Tsuji, and I. Takewaki. 2015. "Innovative base-isolated building with large mass-ratio TMD at basement for greater earthquake resilience." *Future Cities Environ.* 1 (1): 1–19.

Hogg, S. 2015. "Seismically resilient building technology: examples of resilient buildings constructed in New Zealand since 2013." In *Proc., 10th Pacific Conf. on Earthquake Engineering Building an Earthquake-Resilient Pacific,* Paper 190: 1–10. Auckland, New Zealand: Australian Earthquake Engineering Society.

Hong, W. K., and H. C. Kim. 2004. "Performance of a multi-story structure with a resilient-friction base isolation system." *Comput. Struct.* 82 (27): 2271–2283.

Hu, S., W. Wang, B. Qu, and M. S. Alam. 2020. "Development and validation test of a novel self-centering energy-absorbing dual rocking core (SEDRC) system for seismic resilience." *Eng. Struct.* 211: 110424.

Iemura, H., and T. Taghikhany. 2004. "Optimum design of resilient sliding isolation system to protect equipments." Paper No. 1362. In *Proc., 13th World Conf. on Earthquake Engineering*, Paper 1362. Vancouver: Canadian Association for Earthquake Engineering.

Iemura, H., T. Taghikhany, and S. K. Jain. 2007. "Optimum design of resilient sliding isolation system for seismic protection of equipments." *Bull. Earthq. Eng.* 5 (1): 85–103.

Iemura, H., T. Taghikhany, Y. Takahashi, and S. K. Jain. 2005. "Effect of variation of normal force on seismic performance of resilient sliding isolation systems in highway bridges." *Earthq. Eng. Struct. Dyn.* 34 (15): 1777–1797.

Kamaludin, P. N. C., M. M. Kassem, E. N. Farsangi, F. M. Nazri, and E. Yamaguchi. 2020. "Seismic resilience evaluation of RC-MRFs equipped with passive damping devices." *Earthq. Struct.* 18 (3): 391–405.

Lei, K. M., and A. G. Hemried. 1992. "Seismic response of equipment in resilient-friction base isolated structures." In *Proc., 10th World Conf. on Earthquake Engineering*, 2013–2018. London: Routledge.

Mayes, R. L., A. G. Brown, and D. Pietra. 2012. "Using seismic isolation and energy dissipation to create earthquake-resilient buildings." *Bull. N. Z. Soc. Earthq. Eng.* 45(3): 117–122.

Mokhtari, M., and H. Naderpour. 2020. "Seismic resilience evaluation of base-isolated RC buildings using a loss-recovery approach." *Bull. Earthq. Eng.* 18 (10): 5031–5061.

Moretti, S., A. Trozzo, V. Terzic, G. P. Cimellaro, and S. Mahin. 2014. "Utilizing base-isolation systems to increase earthquake resiliency of healthcare and school buildings." *Procedia Econ. Fin.* 18: 969–976.

Mostaghel, N., and A. R. Mortazavi. 1991. "An assessment of SEAONC draft code for resilient sliding isolators." *Earthq. Eng. Struct. Dyn.* 20 (6): 523–533.

Oh, J., G. H. Hyeon, Y. S. Park, and S. K. Park. 2008. "A study on aseismatic performance of base isolation systems using resilient friction pot bearing." *J. Korea Inst. Struct. Maint. Inspection* 12 (1): 127–134.

Rabiee, R., and Y. Chae. 2018. "Adaptive base isolation system to achieve structural resiliency under both short- and long-period earthquake ground motions." *J. Intell. Mater. Syst. Struct.* 30 (1): 16–31.

Sani, H. P., M. Gholhaki, and M. Banazadeh. 2018. "Simplified direct loss measure for seismic isolated steel moment-resisting structures." *J. Constr. Steel Res.* 147: 313–323.

Stahl, D. C., R. W. Wolfe, and M. Begel. 2004. "Improved analysis of timber rivet connections." *J. Struct. Eng.* 130 (8): 1272–1279.

Takewaki, I., K. Fujita, K. Yamamoto, and H. Takabatake. 2011. "Smart passive damper control for greater building earthquake resilience in sustainable cities." *Sust. Cities Soc.* 1 (1): 3–15.

Tang, Z., G. Clifton, J. Lim, J. Maguire, and L. Teh. 2017. "Increasing seismic resilience of pallet racking systems using sliding friction baseplates." In *2017 NZSEE Conf.*, 1–8. Wellington, New Zealand: New Zealand Society for Earthquake Engineering.

Taniguchi, M., K. Fujita, M. Tsuji, and I. Takewaki. 2016. "Hybrid control system for greater resilience using multiple isolation and building connection." *Front. Built Environ.* 2: 26.

Tong, F., and C. Christopoulos. 2020. "Uncoupled rocking and shear base-mechanisms for resilient reinforced concrete high-rise buildings." *Earthq. Eng. Struct. Dyn.* 49 (10): 981–1006.

Venkittaraman, A., and S. Banerjee, 2014. "Enhancing resilience of highway bridges through seismic retrofit." *Earthq. Eng. Struct. Dyn.* 43 (8): 1173–1191.

Wada, A. 2010. "Seismic design for resilient society." Keynote lecture. In *Proc., Joint 7th Int. Conf. Urban Earthquake Engineering. and 5th Int. Conf. on Earthquake Engineering*, 1–7. Tokyo: Tokyo Institute of Technology.

Wang, W., C. Fang, A. Zhang, and X. Liu. 2019. "Manufacturing and performance of a novel self-centering damper with shape memory alloy ring springs for seismic resilience." *Struct. Control Health Monit.* 26 (5): e2337.

REFERENCES

ACI (American Concrete Institute). 2007. *Acceptance criteria for special unbonded post-tensioned precast structural walls based on validation testing and commentary*. ACI Innovation Task Group 5.1. Farmington Hills, MI: ACI.

ACI. 2009. *Requirements for design of a special unbonded post-tensioned precast shear wall satisfying ACI ITG-5.1 and commentary.* ACI ITG-5.2. ACI Innovation Task Group 5.2. Farmington Hills, MI: ACI.

ACI. 2019a. *Acceptance criteria for special unbonded post-tensioned precast structural walls based on validation testing (ACI 550.6-19) and commentary (ACI 550.6R-19).* ACI Innovation Task Group 5. Farmington Hills, MI: ACI.

ACI. 2019b. *Requirements for design of a special unbonded post-tensioned precast shear wall satisfying ACI 550.6 (ACI 550.7-19) and commentary (ACI 550.7R-19).* ACI Innovation Task Group 5. Farmington Hills, MI: ACI.

Agrawal, A. K., and J. N. Yang. 2000. "A semi-active electromagnetic friction damper for response control of structures." In *Advanced technology in structural engineering*, M. Elgaaly, ed., 1–8. New York: ASCE.

Aiken, L. D., and J. M. Kelly. 1990. *Earthquake simulator testing and analytical studies of two energy-absorbing systems for multistory structures.* UCB/EERC-90/03. Berkeley, CA: Earthquake Engineering Research Center.

Aiken, L. D., D. K. Nims, and J. M. Kelly. 1992. "Comparative study of four passive energy dissipation system." *Bull. N. Z. Natl. Soc. Earthquake Eng.* 25 (3): 175–192.

Allmond, J. D., and B. L. Kutter. 2014. "Design considerations for rocking foundations on unattached piles." *J. Geotech. Geoenviron. Eng.* 140 (10): 4014058.

Antonellis, G., and M. Panagiotou. 2014. "Seismic response of bridges with rocking foundations compared to fixed-base bridges at a near-fault site." *J. Bridge Eng.* 19 (5): 04014007.

ASCE. 2017. *Minimum design loads for buildings and other structures.* ASCE 7-16. Reston, VA: ASCE.

Bozorgnia, Y., and V. V. Bertero. 2004. *Earthquake engineering.* Boca Raton, FL: CRC Press.

Bruneau, M., S. E. Chang, R. T. Eguchi, G. C. Lee, T. D. O'Rourke, A. M. Reinhorn, et al. 2003. "A framework to quantitatively assess and enhance the seismic resilience of communities." *Earthquake Spectra* 19 (4): 737–738.

Buchanan, A., B. Deam, M. Fragiacomo, S. Pampanin, and A. Palermo. 2008 "Multi-storey prestressed timber buildings in New Zealand." *Struct. Eng. Int.* 18 (2): 166–173.

Buckle, I. G. 1989. "Recent development in isolated hardware." In *Proc., ASCE Cong. '89 on Seismic Engineering and Structures,* 789–798. Reston, VA: ASCE.

Buckle, I. G., and R. L. Mayes. 1990. "Seismic isolation: History, application, and performance—A world view." *Earthquake Spectra* 6 (2): 161–201.

Chang, C., A. Pall, and J. Louie. 2006. "The use of friction dampers for seismic retrofit of the Monterey County Government Center." In *Proc., 8th US National Conference on Earthquake Engineering,* San Francisco, April, Paper 951. Oakland, CA: EERI.

Colajanni, P., and M. Papia. 1997. "Hysteretic behavior characterization of friction-damped braced frames." *J. Struct. Eng.* 123 (8): 1020–1028.

Constantinou, M. C., T. T. Soong, and G. F. Dargush. 1998. *Passive energy dissipation systems for structural design and retrofit.* Buffalo, NY: University at Buffalo, Multidisciplinary Center for Earthquake Engineering Research.

Couch, L. 2020. "Resilience and response of the dual system braced frame with friction damper." Master's thesis, California State University, Fresno, Dept. Civil and Geomatics Engineering.

Danner, M., and C. G. Clifton. 1995. *Development of moment-resisting steel frames incorporating semi-rigid elastic joints.* 1994/95 Research Rep., HERA Rep. R4-87. Auckland, New Zealand: HERA.

Englekirk, R. E. 2002. "Design-construction of the paramount—A 39-story precast prestressed concrete apartment building." *PCI J.* 47 (4): 56–71.

Espinoza, A., and S. Mahin. 2006. "Rocking of bridge piers subjected to multi-directional earthquake loading." In *Proc., 8th National Conf. on Earthquake Engineering*, Paper 965, 1–7. San Francisco: EERI.

Ettouney, M. 2014. *Resilience management: How it is becoming essential to civil infrastructure recovery*. New York: McGraw Hill.

Fan, F. G., G. Ahmadi, N. Mostaghel, and I. G. Tadjbakhsh. 1991. "Performance analysis of aseismic base isolation systems for a multi-story building." *Soil Dyn. Earthquake Eng.* 10 (3): 152–171.

Fan, F. G., G. Ahmadi, and I. G. Tadjbakhsh. 1990a. "Multi-story base-isolated buildings under a harmonic ground motion—Part I: A comparison of performances of various systems." *Nucl. Eng. Des.* 123 (1): 1–16.

Fan, F. G., G. Ahmadi, and I. G. Tadjbakhsh. 1990b. "Multi-story base-isolated buildings under a harmonic ground motion—Part II: Sensitivity analysis." *Nucl. Eng. Des.* 123 (1): 17–26.

Feng, M. Q., M. Shinozuka, and S. Fujii. 1993. "Friction controllable sliding isolation system." *J. Eng. Mech.* 9: 1845–1864.

Ferritto, J. M. 1991. "Studies on seismic isolation of buildings." *J. Struct. Eng.* 117 (11): 3293–3314.

Filiatrault, A., and T. Sullivan. 2014. "Performance-based seismic design of nonstructural building components: The next frontier of earthquake engineering." *Earthquake Eng. Eng. Vibr.* 13: 17–46.

FitzGerald, T. F., T. Anagnos, M. Goodson, and T. Zsutty. 1989. "Slotted bolted connections in aseismic design for concentrically braced connections." *Earthquake Spectra* 5 (2): 383–391.

Fleischman, R. B., C. Naito, J. Restrepo, R. Sause, S. K. Ghosh, G. Wan, et al. 2005. "Precast diaphragm seismic design methodology (DSDM) project, part 2: research program." *PCI J.* 50 (6): 14–31.

Gilmore, A. T. 2012. "Options for sustainable earthquake-resistant design of concrete and steel buildings." *Earthquake Struct.* 3 (6): 783–804.

Golafshani, A. A., H. R. Mirdamadi, and F. M. Tehrani. 1996. *Application of base isolation systems in low-rise buildings*. [In Persian.] Technical Rep. Research Institute. Tehran, Iran: Sharif University of Technology.

Grigorian, M., and C. E. Grigorian. 2018. "Sustainable earthquake-resisting system." *J. Struct. Eng.* 144 (2): 04017199.

Haider, J. R., and U. Kim. 2012. "A parametric approach for the optimization of passive friction dampers." In *Structures Congress*, edited by J. Carrato and J. Burns, 1673–1684. Chicago. Reston, VA: ASCE.

Hamburger, R. O., and W. T. Holmes. 1998. *Vision statement: EERI/FEMA performance-based seismic engineering project, background document for the EERI/FEMA action plan*. Oakland, CA: Earthquake Engineering Research Institute.

Harden, C. W., and T. C. Hutchinson. 2009. "Beam-on-nonlinear-Winkler-foundation modeling of shallow, rocking-dominated footings." *Earthquake Spectra* 25: 277–300.

Hassani, A., F. M. Tehrani, and S. Nasehpur. 2001. *Rubber bearings for earthquake protection of buildings*. [In Persian.] Tehran, Iran: Building and Housing Research Center.

Hassani, A., F. M. Tehrani, and M. Tehranizadeh. 1996. *Sliding foundation system for small aseismic buildings*. [In Persian.] Tehran, Iran: Building and Housing Research Center.

Henry, R. S. 2011. "Self-centering precast concrete walls for buildings in regions with low to high seismicity." Ph.D. thesis, University of Auckland, Dept. Civil and Environmental Engineering.

Holden, T. J. 2001. *A comparison of the seismic performance of precast wall construction: Emulation and hybrid approaches*. Research Rep. 2001-04. Christchurch, New Zealand: University of Canterbury.

Housner, G. 1963. "The behavior of inverted pendulum structures during earthquakes." *Bull. Seismol. Soc. Am.* 53 (2): 403–417.

Jangid, R. S., and J. M. Kelly. 2001. "Base isolation for near-fault motions." *Earthquake Eng. Struct. Dyn.* 30 (5): 691–707.

Kanno, Y., S. Fujita, and Y. Ben-Haim. 2017. "Structural design for earthquake resilience: Info-gap management of uncertainty." *Struct. Saf.* 69: 23–33.

Kasai, K., Y. Ooki, S. Motoyui, T. Takeuchi, and E. Sato. 2007. "E-defense tests on full-scale steel buildings: Part 1—Experiments using dampers and isolators." In *Research Frontiers at Structures Congress*, edited by K. Kasai, Y. Ooki, S. Motoyui, and T. Takeuchi, 1–12. Reston, VA: ASCE.

Kawashima, K., and T. Nagai. 2006. "Effectiveness of rocking seismic isolation on bridges." In *Proc., 4th Int. Conf. on Earthquake Engineering*, Paper 086. Taipei, Taiwan: National Center for Research on Earthquake Engineering.

Kelly J. M. 1990. "Base isolation: Linear theory and design." *Earthquake Spectra* 6 (2): 223–244.

Kelly, J. M. 1999. "The role of damping in seismic isolation." *Earthquake Eng. Struct. Dyn.* 28 (1): 3–20.

Kiran, A. R., M. K. Agrawal, and G. R. Reddy. 2016. "Seismic retrofitting of a process column using friction dampers." *Procedia Eng.* 144: 1356–1363.

Kobori, T., M. Takahashi, T. Nasu, N. Niwa, and K. Ogasawara. 1993. "Seismic response controlled structure with active variable stiffness system." *Earthquake Eng. Struct. Dyn.* 22 (11): 925–941.

Kurama, Y., R. Sause, S. Pessiki, and L. W. Lu. 1999. "Lateral load behavior and seismic design of unbonded post-tensioned precast concrete walls." *ACI Struct. J.* 96 (4): 622–632.

Li, L. 1984. "Base isolation measure for aseismic building in China." In Vol. 6 of *Proc., 8th World Conf. on Earthquake Engineering*, 791–798. Tokyo, Japan: International Association for Earthquake Engineering (IAEE).

Lin, Y. C., J. Ricles, and R. Sause. 2009. "Earthquake simulations on a self-centering steel moment resisting frame with web friction devices." In *Proc., 2009 Structures Congress,* 1–10. Reston, VA: ASCE.

Liu, Q. 2016. "Study on interaction between rocking-wall system and surrounding structure." Ph.D. thesis, University of Minnesota, Dept. of Civil, Environmental, and Geo-Engineering.

Lou, J. Y. K., L. D. Lutes, and J. J. Li. 1994. "Active tuned liquid damper for structural control." In *Proc., 1st World Conf. on Structural Control,* edited by G. W. Housner, B. A. Klink, and E. Kausel, 70–79. Oxford, UK: Elsevier.

MacRae G. A. 2010. "Some steel seismic research issues." In *Proc., Steel Structures Workshop 2010, Research Directions for Steel Structures,* compiled by G. A. MacRae and G. C. Clifton. Christchurch, New Zealand: University of Canterbury.

Mahin, S. 2008. "Sustainable design considerations in earthquake engineering." In *Proc., 14th World Conf. on Earthquake Engineering,* 1–8. Tokyo, Japan: International Association for Earthquake Engineering.

Mahin, S. 2017. "Resilience by design: A structural engineering perspective." In *Proc., 16th World Conf. on Earthquake Engineering,* Keynote Lecture, 9–13. Santiago, Chile: Chilean Association on Seismology and Earthquake Engineering.

Marriot, D. 2009. "The development of high-performance post-tensioned rocking systems for the seismic design of structure." Ph.D. thesis, University of Canterbury, Dept. of Civil Engineering.

Mayes, R. L., A. G. Brown, and D. Pietra. 2012. "Using seismic isolation and energy dissipation to create earthquake-resilient buildings." *Bull. New Zealand Soc. Earthq. Eng.* 45 (3): 117–122.

Mayes, R. L., I. G. Buckle, and L. R. Jones. 1988. "Seismic isolation—Solution to the earthquake problem of the precast concrete industry." *PCI J.* 33 (3): 24–57.

McMahon, S., and N. Makris. 1997. "Large-scale ER-damper for seismic protection of bridges." In *Proc., Structures Congress,* 1451–1455. Reston, VA: ASCE.

Morgan, T., A. 2007. "The use of innovative base isolation systems to achieve complex seismic performance objectives." Ph.D. thesis, University of California, Dept. Civil and Environmental Engineering.

Mostaghel, N. 1984. *Resilient-friction base isolator.* Rep. No. UTEC 84-097. Salt Lake City, UT: University of Utah.

Naeim, F. 2001. *The seismic design handbook.* New York, NY: Springer.

Naeim, F., and J. M. Kelly. 1999. *Design of seismic isolated structures.* New York: Wiley.

Nazari, M. 2016. "Seismic performance of unbonded post-tensioned precast wall systems subjected to shake table testing." Ph.D. thesis, Iowa State University, Dept. of Civil, Construction, and Environmental Engineering.

Nazari, M., and S. Sritharan. 2018. "Dynamic evaluation of PreWEC systems with varying hysteretic energy dissipation." *J. Struct. Eng.* 144 (10): 04018185.

Nazari, M., and S. Sritharan. 2019. "Seismic design of precast concrete rocking wall systems with varying hysteretic damping." *PCI J.* 64 (5): 58–76.

Nazari, M., and S. Sritharan. 2020. "Influence of different damping components on dynamic response of concrete rocking walls." *J. Eng. Struct.* 212: 110468.

Nazari, M., S. Sritharan, and S. Aaleti. 2017. "Single precast concrete rocking walls as earthquake force-resisting elements." *Earthquake Eng. Struct. Dyn.* 46 (5): 753–769.

Negro, P., G. Verzeletti, J. Molina, S. Pedretti, D. Lo Presti, and S. Pedroni. 1998. *"Large-scale geotechnical experiments on soil–foundation interaction."* Special Pub. No. I.98.73. Ispra, Italy: European Commission, Joint Research Center.

Naghshineh, A., A. Kassem, A. G. Pilorge, O. R. Galindo, and A. Bagchi. 2018. "Seismic performance of reinforced concrete frame buildings equipped with friction dampers." In *Proc., Structures Congress 2018*, 94–101. Reston, VA: ASCE.

Nims, D. K., P. J. Richter, and R. E. Bachman. 1993. "Use of the energy dissipating restraint for seismic hazard mitigation." *Earthquake Spectra* 9 (3): 476–489.

NRCC (National Research Council Canada). 2005. *National building code of Canada*. Ottawa: NRCC.

NZS (Standards New Zealand). 2006. *Concrete structures standards*. NZS 3101-06. Wellington, New Zealand: Standards New Zealand.

Palermo, A., S. Pampanin, A. H. Buchanan, and M. P. Newcombe. 2005. "Seismic design of multi-storey buildings using laminated veneer lumber (LVL)." In *Proc., New Zealand Society of Earthquake Engineering, Annual Conf.*, Paper 14: 1-8. Christchurch, New Zealand: University of Canterbury.

Pall, A. S. 1979. "Limited slip bolted joints a device to control the seismic response of large panel structures." Ph.D. thesis, Concordia University, The Faculty of Engineering.

Pall, A. S. 1984. "Response of friction damped buildings." In Vol. V of *Proc., 8th World Conf. on Earthquake Engineering*, 1007–1014. Tokyo, Japan: International Association for Earthquake Engineering.

Pall, A. S., and C. Marsh. 1982. "Response of friction damped braced frames." *J. Struct. Eng.* 108 (ST6): 1–13.

Pall, A. S., C. Marsh, and P. Fazio. 1980. "Friction joints for seismic control of large panel structures." In Vol. 3 of *Proc., AICAP/CEB Seismic Conf., Rome, Italy*, 27–34. Paris, France: Comite Euro–International Du Beton.

Pall, A. S., and R. Pall. 1996. "Friction dampers for seismic control of buildings a Canadian experience." In *Proc., 11th World Conf. on Earthquake*

Engineering, Acapulco, Mexico, Paper 497. Pedregal, Tlalpan: Sociedad Mexicana de Ingeniería Sísmica.

Pampanin S. 2015. "Towards the 'ultimate earthquake-proof' building: development of an integrated low-damage system." In Vol. 39 of *Perspectives on European earthquake engineering and seismology. geotechnical, geological and earthquake engineering*, A. Ansal, ed., 321–358. Cham, Switzerland: Springer.

Pampanin, S., W. Kam, G. Haverland, and S. Gardiner. 2011. "Expectation meets reality: Seismic performance of post-tensioned precast concrete southern cross endoscopy building during the 22nd Feb 2011 Christchurch earthquake." In *NZ Concrete Industry Conf.*, 1–14. Wellington, New Zealand: Concrete NZ.

Patil, C. S., V. B. Patil, and S. B. Kharmale. 2015. "Performance-based plastic design of a high rise moment resisting frame with friction dampers." *Earthq. Resistant Eng. Struct.* 152: 73–82.

Pessiki, S. 2017. "Sustainable seismic design." *Procedia Eng.* 171: 33–39.

Priestley, M., and J. Tao. 1993. "Seismic response of precast prestressed concrete frames with partially debonded tendons." *PCI J.* 38 (1): 58–69.

Priestley, M. J. N., S. Sritharan, J. R. Conley, and S. Pampanin. 1999. "Preliminary results and conclusions from the PRESSS five-story precast concrete test building." *PCI J.* 44 (6): 42–67.

Rahman, A. M., and J. I. Restrepo. 2000. *Earthquake resistant precast concrete buildings: Seismic performance of cantilever walls prestressed using unbonded tendons*. Research Rep. 2000-5. Christchurch, New Zealand.

Roik, K., U. Dorka, and P. Dechent. 1988. "Vibration control of structures under earthquake loading by three-stage friction-grip elements." *Earthq. Eng. Struct. Dyn.* 16: 501–521.

Roke, D., R. Sause, J. M. Ricles, C. Y. Seo, and K. S. Lee. 2006. "Self-centering seismic-resistant steel concentrically-braced frames." In *Proc., 8th US National Conference on Earthquake Engineering*, 5845–5854. Oakland, CA: Earthquake Engineering Research Institute.

Saad, A., D. H. Sanders, and I. G. Buckle. 2012. "Impact of rocking foundations on horizontally curved bridge systems subjected to seismic loading." In *Structures Congress*, edited by J. Carrato and J. Burns, 625–635. Reston, VA: ASCE.

Saatcioglu, M., D. Palermo, A. Ghobarah, D. Mitchell, R. Simpson, P. Adebar, et al. 2012. "Performance of reinforced concrete buildings during the 27 February 2010 Maule (Chile) earthquake." *Can. J. Civ. Eng.* 40 (8): 693–710.

SEAOC (Structural Engineers Association of California). 1995. *Vision 2000: A framework for performance-based engineering*. Sacramento, CA: SEAOC.

Sidwell G. 2010. "The Te Puni apartment buildings." In *Proc., Steel Structures Workshop 2010, Research Directions for Steel Structures*, compiled by G. A. MacRae and G. C. Clifton. Christchurch, New Zealand: University of Canterbury.

Sritharan, S., S. Aaleti, R. S. Henry, K. Y. Liu, and K. C. Tsai. 2008. "Introduction to PreWEC and key results of a proof of concept test." In *M. J. Nigel Priestley Symp.*, 95–106. Pavia, Italy: IUSS Press.

Sritharan, S., S. Aaleti, R. Henry, K. Liu, and K. Tsai. 2015. "Precast concrete wall with end columns (PreWEC) for earthquake resistant design." *Earthq. Eng. Struct. Dyn.* 44 (12): 2075–2092.

Stanton, J. F., and S. D. Nakaki. 2002. *Design guidelines for precast concrete seismic structural systems unbonded post-tensioned split walls*. PRESSS Rep. No. 01/03-09. Seattle: University of Washington.

Su, L., G. Ahmadi, and I. G. Tadjbakhsh. 1989. "A comparative study of various base isolation systems, Part I: Shear beam structures." *Earthq. Eng. Struct. Dyn.* 18 (1): 11–32.

Su, L., G. Ahmadi, and I. G. Tadjbakhsh. 1990. "A comparative study of various base isolation systems, Part II: sensitivity analysis." *Earthq. Eng. Struct. Dyn.* 19 (1): 21–33.

Su, L., G. Ahmadi, and I. G. Tadjbakhsh. 1991. "Performance of sliding resilient-friction base-isolation system." *J. Struct. Eng.* 117 (1): 165–181.

Sullivan, T. J., D. P. Welch, and G. M. Calvi. 2014. "Simplified seismic performance assessment and implications for seismic design." *Earthq. Eng. Eng. Vibr.* 13: 95–122.

Symans, M. D., and M. C. Constantinou. 1997. "Semi-active control systems for seismic protection of structures: a state of the art review." *Eng. Struct.* 21 (6): 469–487.

Takewaki, I., A. Moustfa, and K. Fujita. 2003. *Improving the earthquake resilience of buildings*. London, UK: Springer.

Tehrani, F. M. 1990. "Review of Manjīl 1990 Earthquake damages of Buildings in Gīlān and Zanjān." [In Persian.] *Omrān, Magazine of Civil Engineering Students of the Sharif University of Technology* 2: 5–7.

Tehrani, F. M. 1992. *Base isolation systems*. [In Persian.] Master's Seminar Report. Amirkabir University of Technology, Dept. Civil Engineering.

Tehrani, F. M. 1993a. "Base isolation systems." [In Persian.] In *Proc., 2nd Workshop on Natural Disasters*, 26–29. Shiraz, Iran: Shiraz University.

Tehrani, F. M. 1993b. "Sliding base isolation system for unreinforced masonry buildings." *Master's thesis*, Amirkabir University of Technology, Dept. Civil Engineering.

Tehrani, F. M. 1993c. "Studies on base isolation devices." [In Persian.] *Omrān, Magazine of Civil Engineering Students of the Sharif University of Technology* 11: 10–12.

Tehrani, F. M. 1994. "Introducing a sliding foundation system for small buildings in Iran." [In Persian.] In *Proc., 2nd Conf. of Civil Engineering Students*, Paper 37. Tehran: Tehran University.

Tehrani, F. M. 1996. "Comparative evaluation of dynamic behavior of base isolated structures." [In Persian.] In *Proc., 3rd National Seminar on Analytical and Experimental Study of Dynamic Response of Structures*, 73–83. Gīlān, Iran: Gīlān University.

Tehrani, F. M. 1998. *Sliding bearing systems and material characteristics for sliding layers*. [In Persian.] Tehran: National Disasters Prevention Center of Iran.

Tehrani, F. M. 2000. *Sliding foundation system with safety margin*. IR PRO Patent 26443, filed December 31, 1996, and issued March 20, 2000.

Tehrani, F. M., and G. Ahmadi. 2015. "Pure-friction base-isolation system for masonry buildings: A case study from research to construction." In *Proc., Engineering Mechanics Institute Conf*, Paper 891. Stanford, CA: Stanford University.

Tehrani, F. M., and A. Hassani. 1994a. "Isolating foundations." [In Persian.] *Sarpanāh, Magazine of Ministry of Housing and Urbanizing* 92: 17–20.

Tehrani, F. M., and A. Hassani. 1994b. "Stability of small base isolated buildings on sliding foundations." [In Persian.] In *Proc., 1st Seminar of Housing Development Policies*, 499–515. Tehran: Ministry of Housing and Urbanizing.

Tehrani, F. M., and A. Hassani. 1995. "Stability of seismically isolated masonry buildings against earthquake." [In Persian.] In *Proc., 2nd Int. Conf. on Seismology and Earthquake Engineering*, 1169–1177. Tehran: International Institute of Seismology and Earthquake Engineering.

Tehrani, F. M., and A. Hassani. 1996. "Behavior of Iranian low rise buildings on sliding base subjected to earthquake excitation." In *Proc., 11th World Conf. on Earthquake Engineering*. Paper 1433: 1–9. Oxford, UK: Elsevier.

Tehrani, F. M., and A. Massoud. 1995. "Experimental studies of dynamic behavior of structures on sliding foundation subjected to sinusoidal excitation." [In Persian.] In *Proc., 2nd National Seminar on Analytical and Experimental Study of Dynamic Response of Structures*, 81–98. Gīlān, Iran: Gīlān University.

Tehrani, F. M., M. Tehranizadeh, and A. Hassani. 1993. "Sliding foundation system for base isolated small buildings." [In Persian.] In *Proc., 8th Int. Seminar on Earthquake Prognostics*, 176–191. Tehran: Iranian Center for Studies on Natural Disasters.

Tehrani, F. M., M. Tehranizadeh, and A. Hassani. 1994. "Analytical behavior of sliding base isolators." [In Persian.] In *Proc., 1st National Seminar on Analytical and Experimental Study of Dynamic Response of Structures*, 60–74. Gīlān, Iran: Gīlān University.

Tehranizadeh, M. and F. M. Tehrani. 1997. "Sliding base isolation system for unreinforced small masonry bearing wall buildings." [In Persian.] *Amirkabir J. Sci. Res.* 19 (2): 166–176.

Turek, M., C. E. Ventura, and K. Thibert. 2008. "Dynamic characteristics of tall buildings in Vancouver." *J. Struct. Eng.* 135 (8): 883–886.

Twigden, K. M. 2016. "Dynamic response of unbonded post-tensioned concrete walls for seismic resilient structures." Ph.D. dissertation, University of Auckland, Dept. Civil and Environmental Engineering.

Vail, C., and J. Hubbell. 2003. "Structural upgrade of the Boeing commercial airplane factory at Everett, Washington." In *Proc., Structures Congress & Exposition*, Paper 000529. Reston, VA: ASCE.

Vugrin, E. D., D. E. Warren, M. A. Ehlen, and R. C. Camphouse. 2010. "A framework for assessing the resilience of infrastructure and economic systems." In *Sustainable and resilient critical infrastructure systems*, K. Gopalakrishnan and S. Peeta, eds., 77–116. Berlin: Springer.

Yousuf, M., and A. Bagchi. 2009. "Seismic design and performance evaluation of steel-frame buildings designed using the 2005 National building code of Canada." *Can. J. Civ. Eng.* 36: 280–294.

Zayas, V., S. Low, and S. Mahin. 1989. "Shake table testing of a friction pendulum seismic isolation system." In *Seismic Engineering: Research and Practice*, edited by C. A. Kircher and A. K. Chopra. New York: ASCE.

CHAPTER 7

RECOMMENDATIONS ON ACHIEVING HEALTHCARE RESILIENCE FOLLOWING EXTREME EVENTS

Hussam Mahmoud, Emad M. Hassan

7.1 INTRODUCTION

The devastating losses that result from earthquakes have attracted vast research attention over the years. The focus has primarily been placed on achieving certain performance objectives such as life safety and collapse preventions that are associated with certain drift limit states (ASCE 2010). Other studies and guidelines have been focused primarily on maintaining a certain level of functionality for various infrastructures after earthquakes and ensuring a quick recovery for the main communities' infrastructure (FEMA 2007, 2010). Ensuring the continuation of vital community services such as hospitalization is specifically critical for minimizing social losses after extreme events (Hassan and Mahmoud 2020). Shortage of hospitalization services could have catastrophic short-term and long-term effects on the community (Hassan and Mahmoud 2020, TariVerdi et al. 2019). This could include, for example, an increase in morbidity and mortality (Sprivulis et al. 2006) that results from direct injuries as well as population outmigration and social instability, during community recovery, eventually leading to other cascading social and economic consequences.

Maintaining hospitalization services is an essential community resilience goal and is vital for ensuring the social stability of any modern community (NIST 2016). Despite the importance of the healthcare sector, structures hosting healthcare facilities usually experience extensive damage and loss of functionality after major earthquakes. For instance, during the 1994 Northridge earthquake, in California, 11 hospitals were damaged and 8 acute care hospitals had to evacuate at least one inpatient (Schultz et al. 2003). The 2003 Bam earthquake, in the Kerman province in

Iran, destroyed almost all healthcare systems, causing 12,000 patients to be taken to hospitals in other provinces (UNICEF 2004). The 2005 Kashmir earthquake, in the Kashmir region, disturbed 68% of the healthcare facilities in the area and most of the severely injured patients had to be evacuated (IASC 2005). The 2009 L'Aquila earthquake, in the region of Abruzzo in central Italy, also resulted in damage to the San Salvatore hospital, requiring evacuation of all patients (Petrazzi et al. 2013). During the 2010 Haiti earthquake, 50 hospitals and health centers were damaged (GOH 2010). More recently, the 2015 Nepal earthquake affected 90% of the health institutions in the Nuwakot district (Nielsen et al. 2016), whereas the 2017 Central Mexico earthquake caused major structural damage to 83 healthcare facilities (Alcalde-Castro et al. 2018) and the collapse of the main hospital and patient evacuation (Villegas et al. 2017).

Based on the previously mentioned historical seismic events, it is imperative that a comprehensive investigation into the impact of earthquakes on healthcare systems be conducted so that a clear understanding of the functionality and resilience of hospitals following earthquakes can be made. To achieve this objective, researchers have investigated damage to different healthcare system components, including structural systems (Ferraioli 2015, Hassan and Mahmoud 2017, Korkmaz et al. 2012, Chen et al. 2010, Jennings and Housner 1971) and nonstructural components, fixtures, and equipment (Schultz et al. 2003, Korkmaz et al. 2012, McIntosh et al. 2012, Youance and Nollet 2012, Achour et al. 2011) after major earthquakes. Other studies focused on seismic vulnerability and risk of hospital buildings and their different components (Hassan and Mahmoud 2017, Nikfar and Konstantinidis 2019, Di Sarno et al. 2018, Nikfar and Konstantinidis 2017, Karapetrou et al. 2016, Bilgin 2016, Cosenza et al. 2015). In addition, the expected surge in demand on hospitals after seismic events, which is critical to quantify to understand the impact of such surge on the quality of the offered medical services, was also simulated in many studies (Therrien et al. 2017, Achour et al. 2016, Watson et al. 2013, Stratton and Tyler 2006, McCarthy et al. 2006, Asplin et al. 2006). Functionality and resilience of the healthcare facilities are also addressed by various studies, in which different approaches were presented to quantify different functionality components and resilience metrics of single hospitals and/or hospital clusters (Hassan and Mahmoud 2020; TariVerdi et al. 2019; Perrone et al. 2015; Mulyasari et al. 2013; Miniati and Iasio 2012; McDaniels et al. 2008; Ceferino et al. 2020; Lupoi et al. 2013; Jacques et al. 2014; Hassan and Mahmoud 2018, 2019; Shang et al. 2020; Cimellaro and Pique 2016; Malavisi et al. 2015). Although many of the studies pertained to earthquakes as a hazard, others investigated hospital functionality after extreme weather events (Chand and Loosemore 2016, Masko et al. 2011) or other stressors (Guinet and Faccincani 2016, Vugrin et al. 2015, Hiete et al. 2011, Arboleda 2006).

The functionality of healthcare facilities can be defined as the proportion of services offered by the facility and can be measured using different quantitative and qualitative performance indicators (Hassan and Mahmoud 2019). These performance indicators can, in general, be categorized into quantity and quality components (Cimellaro et al. 2008). The quantity component of the functionality is commonly expressed in terms of the number of staffed beds or the number of patients that can be treated (Hassan and Mahmoud 2020, Lupoi et al. 2013). For proper operation of these beds, in addition to the required trained staff, working space, and adequate supplies (Jacques et al. 2014, Hassan and Mahmoud 2019), it is also essential so that other lifelines are functional including power, water, transportation, telecommunication, among others. Quantifying the quality component of the functionality is a complex process because it depends on different indicators that might be difficult to monitor, measure, or estimate (Cimellaro and Pique 2016). Maxwell (1984) listed six different dimensions of quality of healthcare service, namely, relevance, accessibility, effectiveness, fairness, acceptability and efficiency, and economy. McCarthy et al. (2000) suggested using the patient waiting time before being seen by medical staff to express medical service accessibility, whereas patient treatment time can be utilized to estimate the effectiveness of this offered service (Arboleda et al. 2007). The quality component of functionality also depends on hospital demand and is, therefore, expected to decrease with an increase in the number of casualties (Peek-asa et al. 1998, Lupoi et al. 2013). The resilience of healthcare facilities, on the contrary, can be defined as the capacity of the facility to rebound after a severe disaster such as an earthquake (Cimellaro and Pique 2016). The resilience of healthcare facilities can be quantified as the area underneath the functionality curve according to Cimellaro et al. (2007) and can be described by four dimensions of resilience. Other methods can also be applied to quantify resilience, including graph theory (Berche et al. 2009), fuzzy interference system (Heaslip et al. 2010), compositional demand/supply (Didier et al. 2018), and dynamic finite-element simulations (Mahmoud and Chulahwat 2018), among others.

Understanding the contribution of the healthcare system to the social stability of communities at different recovery stages can aid not only in understanding the main drivers for community losses but also in devising policies for recovery under budget constraints (IOM 2015). The complexity in understanding the interplay between the social stability of communities in the context of healthcare services is further aggravated by the high level of interdependency between the services and other infrastructure in the community such as transportation, power, water, and telecommunication, among others (Cimellaro 2016). Previous studies on the functionality and resilience of healthcare facilities included the interdependence between hospitals and critical infrastructure (Jacques et al. 2014; Hassan and Mahmoud 2018, 2019). For instance, Hiete et al. (2011) investigated the impact of power outage on the healthcare sector, Cimellaro et al. (2013) studied the effect of the ambulance service during extreme events, and Vugrin et al. (2015) studied a hospital's adaptive capacities during the loss of infrastructure services (power, water, and so on). The inclusion of the interdependency between these infrastructures and the investigated hospital, as well as the interaction with other healthcare providers, suppliers, and community individuals, might result in a more complex system that requires a comprehensive model to analyze.

To achieve these objectives, this chapter presents guidance and recommendations for healthcare facilities to increase their functionality and resilience after seismic events that are framed based on previous events and guidelines as well as the analytical analysis conducted using a new framework introduced by Hassan and Mahmoud (2019) to assess the functionality of the healthcare service while accounting for the interdependence among hospitals and among other infrastructure in which losses, functionality, recovery, and resilience are estimated using this framework for a six-story hospital located in Memphis, Tennessee, following an earthquake scenario. Different input parameters are used to estimate hospital resilience, which includes hospital staff, space, and supplies in addition to several decisions that can be made by staff and managers and change the recovery and resilience of the investigated hospital. The analysis results along with the lessons from previous events are used to provide practical recommendations and guidelines for emergency managers and decision makers. These recommendations address both the quantity and quality functionality of the healthcare service and the mutual effect between hospitals within the healthcare and the other lifelines on the healthcare system resilience.

7.2 DEFINITION AND DIMENSIONS OF HOSPITAL FUNCTIONALITY AND RESILIENCE

In this study, we define *hospital resilience* as the normalized area under the total functionality curve from the time of earthquake occurrence to the time of full recovery. *Functionality* depends on the sustained losses and is defined as the ability of the system to offer its intended services. It can be subdivided into different stages that include immediate, short-term, and long-term functionalities. For hospitals, functionality depends on the offered quantity and quality of service, as shown in Figure 7-1. In the event of an earthquake, the quantity portion of hospital functionality can be measured by the number of patients that can be treated or the number of staffed beds available. To operate each bed, different components are required, including personnel, space, and supplies. The quality portion of functionality can be represented in relation to the patients' waiting time at the emergency room

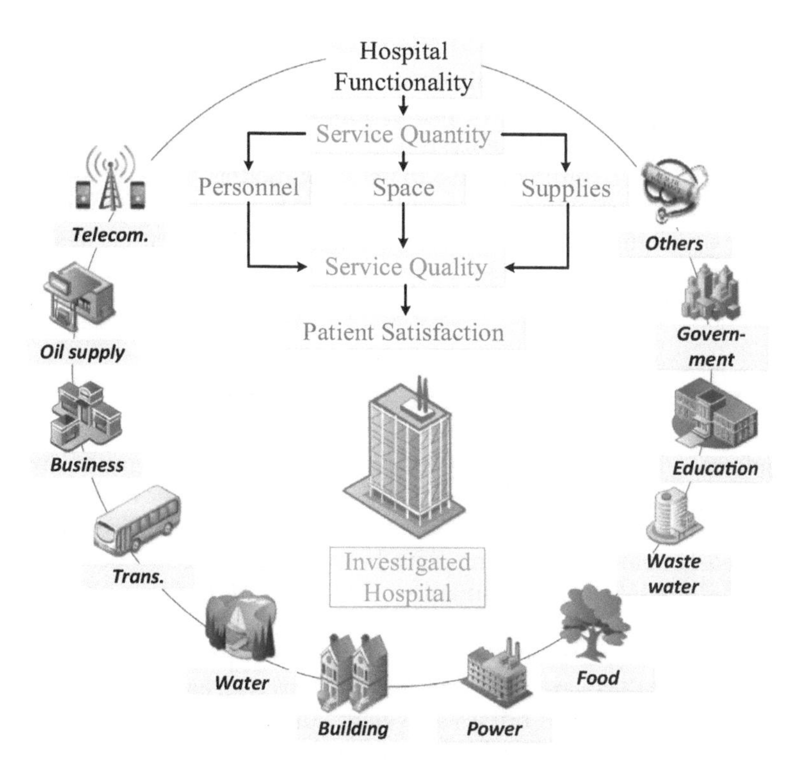

Figure 7-1. Hospital functionality components.

prior to receiving service. It is noteworthy that hospitalization service depends not only on the hospital itself but also on the surrounding lifelines on which the hospital depends. For example, a reduction in transportation network functionality will lead to delays in ambulances' response or even an entire halt to their service. Therefore, while the evaluation of hospital functionality following an extreme event is strongly influenced by damage to the hospital itself, such evaluation should account for the effect of other lifelines, as shown in Figure 7-1. In addition, the inclusion of the interaction between healthcare facilities in the investigation of hospitals' functionality is essential because it can impact the patient demand and assist hospitals to redistribute services and resources.

Immediately after an earthquake, the functionality of healthcare services, as well as other community sectors, is expected to drop because of infrastructure, social, and economic losses resulting from the earthquake damage. Different parameters play critical roles in determining the level of functionality restoration that can be achieved following a major event. These include the type of the damaged component, extent of damage, and available funding resources (e.g., insured losses or federal resources). The recovery process of infrastructure or its components is usually represented by plotting functionality over time. Various studies have investigated the use of different approaches for estimating multiple recovery stages for different lifelines. For example, the statistical curve fitting model (Hazus-MH 2015); expert opinions and regression analysis (ATC 1985); deterministic resource constraint model (Isumi et al. 1985); network models; and Markov chain stochastic models (Kozin and Zhou 1990, Burton et al. 2016, Lin and Wang 2017). To distribute the limited repair resources, different approaches can be utilized by decision makers to direct the recovery process and determine recovery priorities for each community sector.

7.3 HOSPITAL FUNCTIONALITY ASSESSMENT FRAMEWORK

Reduction in hospital functionality can be defined as the ratio between the medical services offered after the hazard to the services offered before the hazard. The functionality itself can be categorized into quality, measured by patients' waiting time before being seen by a healthcare provider (ASCE 2010), and quantity, measured by the number of beds available for patients (FEMA 2007).

The effective operation of these beds requires proper functioning of utilities and medical equipment and supplies as well as the availability of qualified physicians and nurses as well as supporting staff (Jacques et al. 2014). The quantity component of hospital functionality, Q_V, is quantified, in this study, using success-tree analysis (Figure 7-2), comprising the

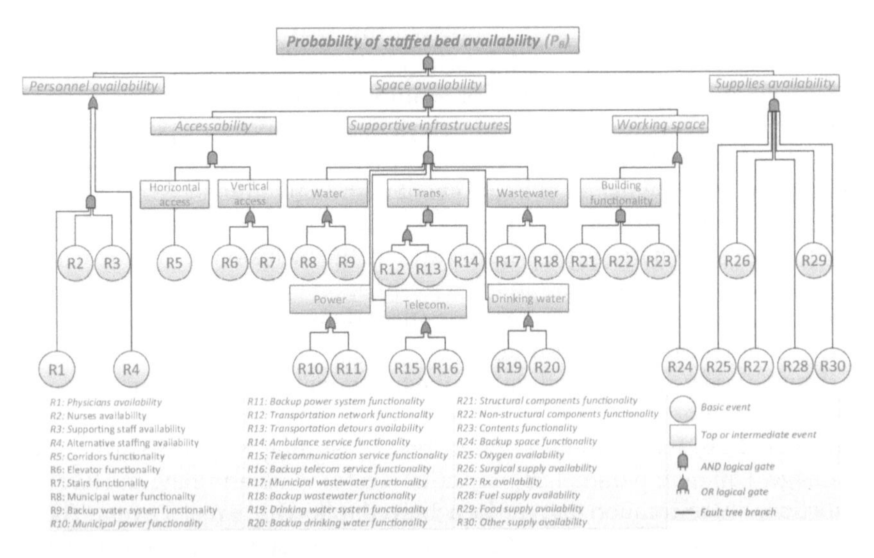

Figure 7-2. Hospital quantitative functionality assessment framework. Source: Hassan and Mahmoud (2019).

quantity of the offered hospitalization service while accounting for the functionality of other lifelines or infrastructure and their interdependence, as shown in Hassan and Mahmoud (2019). The availability of the main events in this success tree is categorized into various intermediate events, including personnel, ST, space, SP, and supplies, SU, in which ST encompasses physicians, nurses, supporting staff, and alternate staff. SP comprises building accessibility and the essential services provided to the healthcare system from the community's infrastructure, which includes water, power, transportation, telecommunication, wastewater, and drinking water as well as structural, nonstructural including equipment, and contents. SU reflects the daily necessities and supplies for the medical facilities such as oxygen, surgical, RX, fuel, and food, among others. Therefore, these basic events not only depend on the hospital building itself but are also highly related to the surrounding community's physical, economic, and social sectors. Details of the success tree and the basic events can be found in Hassan and Mahmoud (2018, 2019). The quality component of hospital functionality, Q_S, represents the patient's satisfaction with the offered hospitalization service. After disasters such as earthquakes, accessibility can be the main dimension controlling hospitalization quality. The waiting time $W(t)$ can be used to express service accessibility according to McCarthy et al. (2000) and includes, in this study, the travel time, basic waiting time, increase owing to shortage of staffed beds, and increase owing to an increased number of patients in the emergency department. Finally, the total functionality is estimated by combining both the quantity and quality functionalities as shown in Equation (7-1):

$$Q = Q_V^{\alpha_V} * Q_S^{\alpha_S} \qquad (7\text{-}1)$$

where a_V and a_S are weighting factors.

A discrete Markov chain process is utilized in this study to estimate recovery for the hospital components separately by discretizing the functionality to different independent sublevels. The effect of different service timelines functionality on the repair process of the hospital building and vice versa is considered in the recovery estimation. A Gantt chart is used to define the repair sequence based on Almufti and Willford (2013), which starts with structural components followed by stairs, elevators, and exterior repairs such as partitions and claddings. The latter can be performed simultaneously with the interior repairs such as the piping, heating, ventilation, and air conditioning, partitions and ceilings, mechanical equipment, and electrical system. The specialization of each repair crew assigned to the different lifelines and the proper repair sequence to eliminate interference between various repair tasks is considered. Recovery of the hospital's supporting lifelines is also calculated using the same approach. Because there are typical limitations in available resources following an extreme event, resources are distributed to different lifelines based on their importance and significance in community recovery using dynamic optimization. Finally, resilience, R, can be graphically estimated as the normalized area under the total functionality curve from the time of earthquake occurrence, t_0, to the total recovery time, t_R, as follows:

$$R_i = \int_{t_0}^{t_R} Q_i(t)dt \tag{7-2}$$

7.4 RECOMMENDATIONS AND GUIDELINES TO IMPROVE RESILIENCE OF HEALTHCARE SYSTEMS

In this section, guidelines for improving the resilience of hospitals are provided based on (a) historical cases and review of existing guidelines and (b) based on a case study of an analyzed hospital using a newly developed hospital functionality framework (Hassan and Mahmoud 2019).

7.4.1 Based on Historical Events and Existing Guidelines

Effect of shortage in hospital staff: Shortage in hospital staff after major earthquakes, or any disaster for that matter, is one of the significant factors contributing to a reduction in healthcare functionality. For instance, a significant reduction in hospital staff was recorded after the 2011 Fukushima earthquake for 18 months (Ochi et al. 2016), and a staff shortage was reported in one-third of the area affected by the 2012 storm Sandy (Levinson 2014). To overcome a shortage in hospital staff, hospitals can borrow staff from other hospitals through voluntary or structured processes. While most of the staff transfer will be voluntary and does not require strict legal agreements, it is recommended to have mutual agreements between hospitals and/or organizations to facilitate staff transfer and work assignments (Paterson et al. 2014). Commonly, healthcare departments define the rules for the mutual aid agreement. Transferring staff can help hospitals accommodate the increase in patient demand and the shortage of hospital staff after an earthquake (IOM 2015). Hospital staff can also work additional hours to cover any staff shortage that might result from death and injuries to some staff members during an earthquake as well as owing to difficulties to reach the hospital because of staff dislocation or damage to the transportation network (Ripp et al. 2012). Some hospitals hire alternate staff to cover shortage in the main staff during normal operation of the hospital. It is worth noting that hospital staff and their families, similar to most of the community residents, suffer after earthquakes and experience higher stress levels. Therefore, providing mental health services for them is recommended so that they can continue to provide the best care for patients.

Effect of backup systems: Interdependency between modern communities' infrastructure is complex as a result of each sector providing services or products to many other sectors. Even though it can increase community productivity, it can also dramatically increase the impact of natural hazards on sectors with higher dependence. Hospitals are dependent on most of the community's lifelines as shown by Cimellaro (2016). Therefore, they can be impacted by the reduction of other lifelines' functionality. As such, it is essential for hospitals to have backup systems and redundant supporting lifelines. Different backup systems can be found in modern hospitals, which include water, power, telecommunication, wastewater, and drinking water systems (Paterson et al. 2014). In FEMA 577 (FEMA 2007), it is suggested that the backup systems are able to support a hospital for at least 4 days after an earthquake. Recommendations to install and maintain hospitals' backup systems are also emphasized in Rodgers et al. (2009). The importance of backup systems after earthquakes was noted following the 2010 Chilean earthquake (Kirsch et al. 2010) and the 2011 Christchurch earthquake (Jacques et al. 2014). It is worth noting that regular maintenance of backup systems and basic training of hospital staff on how to operate them is essential as recorded after Hurricane Sandy (Levinson 2014).

Effect of backup spaces: Hospitals can also include backup spaces, which can be used in cases of shortage in physical beds (Tekin et al. 2017). During the 2017 Central Mexico earthquake (Villegas et al. 2017), the staff evacuated patients to an empty lot that was considered as the backup space. Other hospitals are provided with extra beds that can be used to replace any damaged beds. During disasters, hospitals can add beds in the emergency department and nursing areas, add more patients per room, and convert the lobby into temporary inpatient care (Levinson 2014). However, to operate these extra beds, extra staff and supplies are needed. Some of these beds can be used only as a temporary solution until patients are transferred to another healthcare facility or the hospital is repaired. A sample of the recommendations for establishing and operating hospitals' backup spaces and field hospitals is provided by the World Health Organization (WHO 2003).

Effect of supply availability: After catastrophic disasters such as earthquakes, hospitals as well as other healthcare facilities can suffer a reduction in supplies including medical, fuel, and food supplies (Paterson et al. 2014). This reduction might be a consequence of earthquake damage to the supply rooms and their components (FEMA 2007). The impact on hospital functionality is also expected to amplify because of earthquake damage to the suppliers' facilities and transportation networks (Wang 2014). In addition, the expected increase in patient demand on hospitals after earthquakes will accelerate the consumption of these supplies and add more burden to the healthcare facilities in terms of providing the service. Different approaches can be used by hospitals to solve the supply shortage problem, including, for instance, finding alternate supplies (Paterson et al. 2014), borrowing supplies from other hospitals, finding alternate methods to deliver supplies, and in some cases transferring patients with medical cases that depend on these supplies.

Effect of patients' waiting time: After disasters, hospital staff can work to reduce the patient treatment time, which can eventually elevate the hospital's ability to treat more patients. After the L'Aquila earthquake in 2009, the staff was able to significantly reduce the mean length of stay for patients (Petrazzi et al. 2013). Similarly, at the Canterbury hospital system, after the 2011 Christchurch earthquake, the patient's discharge process was notably accelerated, causing a reduction in hospital bed demand (Jacques et al. 2014). In general, reducing the patients' length of stay, even during normal operation time, is recommended. This could be realized by categorizing a hospital's patients based on their injury severity level. After earthquakes, hospitals can receive two different types of patients (earthquake-related or regular patients). Earthquake severities are classified by HazusMH 2.1 (Hazus-MH 2015) based on their need for medical care into (a) Severity 1, in which no medical care is needed, (b) Severity 2, in which some medical care is needed, (c) Severity 3, where immediate medical care is needed, and (d) Severity 4, which is ultimately death. However, those with Severity 2 and Severity 3 are the only ones considered as a possible hospital patient. Each severity level can also be categorized based on each patient's injury and the required staff and supplies. It is critical to provide communication tools between each hospital staff and ambulance services to identify priorities among the treated patients and availabilities of hospitals within the healthcare system that can receive patients. These tools can effectively reduce overcrowding in emergency departments (Denver Health 2005).

Effect of repair resources: The role played by community decision makers to allocate repair resources within the community is crucial. An obvious plan that clearly identifies community priorities has to be defined before the recovery process starts. By assigning the repair resources, communities can increase infrastructure functionality and enhance both the communities' social and economic fabric. However, not all scenarios can lead to social and economic enhancements. For the healthcare system, maintaining the revenue of hospitals after earthquakes does not necessarily increase the social stability of the community because typically hospitals are not the main drivers for the economy in communities. Therefore, proper distribution and allocation of repair resources among different healthcare facilities are recommended as it has a considerable impact on community stability and resilience.

Effect of patients' transfer: To ensure that a patient will receive the required medical care, the hospital might offer to transfer patients to other hospitals especially when capacity is reached or when treatment is not available (Kulshrestha and Singh 2016). The patient transfer process is a function of the patient case criticality. For instance, some patients can be transferred only using an air ambulance, whereas others can use a regular ambulance, which might complicate the transfer process. Different factors control the patient transfer decision: (1) patient constraint, (2) hospital-to-hospital connection, and (3) receiver hospital availability. The main constraints that might control patient transfer are either case criticality or insurance coverage. On the contrary, the presence of smooth transfer and the availability of connections between hospitals are expected to encourage both the patient and the hospital to make the transfer decision. In addition, the ability of the receiver hospital to safely transfer and treat the patient is expected to have a major effect on this transfer decision. This interaction between healthcare facilities is essential, especially after an earthquake occurrence, as it allows for a redistribution of services (McDaniels et al. 2008). In addition, medical staff shortage can be accommodated among the interacted facilities (Paterson et al. 2014).

7.4.2 Case Study Using a Newly Developed Framework

The considered hospital for the case study is a buckling-restrained braced building, assumed to be located in Memphis, Tennessee, and is designed professionally for the National Earthquake Hazard Reduction Program (NEHRP) (NEHRP Consultants Joint Venture 2013, NEHRP Consultants Joint Venture 2013) in accordance with ASCE 7-10 (ASCE 2010). The building comprises six bays in the N–S direction and five bays in the E–W direction. It is a six-story building with an additional basement floor, as shown in Figure 7-3(a). The span of a given bay is 9.14 m and the floor height is 4.27 m, except for the first floor, serving as a hospital lobby, where the height is 6.10 m. The total area of each floor is 2,506.18 m^2, and the total building height is 31.70 m. Buckling-restrained braces are utilized to resist the lateral loads in both the N–S and the E–W directions. More detailed descriptions of the structure can be found in Hassan and Mahmoud (2018).

With 160 staffed beds, the studied hospital is assumed as a general-purpose healthcare provider with an emergency department and ambulance service. An earthquake intensity, defined by spectral acceleration, S_a, of 1.0 g is assumed to strike the community where the hospital is located at 2:00 a.m. The seismic fragilities were developed, and the direct social and economic losses were estimated by Hassan and Mahmoud (2018) based on this scenario event. Earthquake scenarios in two perpendicular directions were investigated; however, the analysis

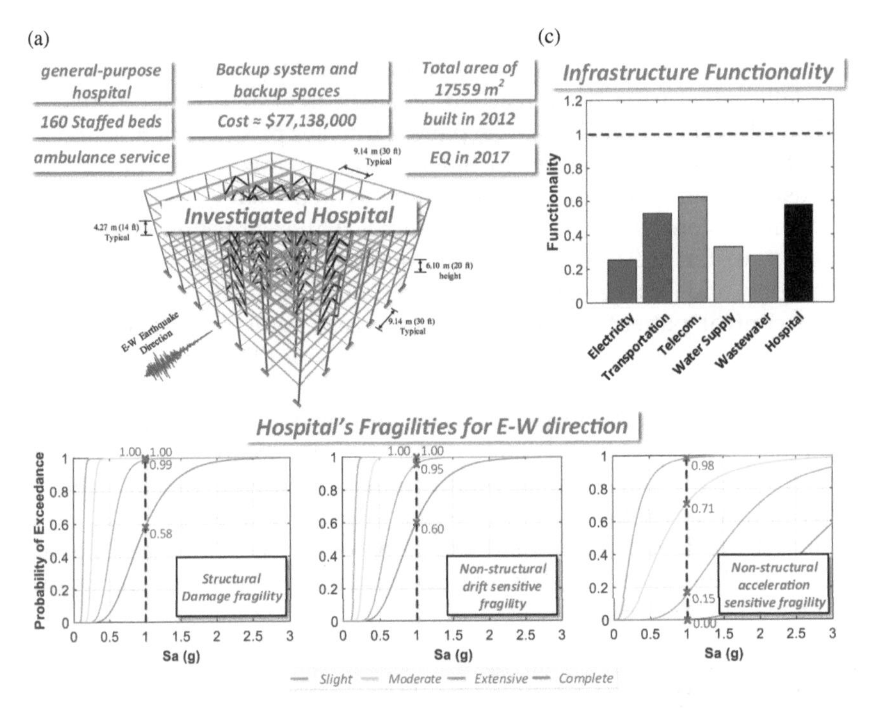

Figure 7-3. (a) Hospital building general configuration, (b) hospital's seismic fragilities, (c) initial drop of functionality for different infrastructures.

showed that the hospital was more vulnerable to earthquakes in the E–W direction. Therefore, in this study, only the earthquake in the E–W direction will be considered. The functionality of the investigated hospital is estimated based on the previously mentioned framework. Based on this scenario earthquake and using the previously mentioned hospital functionality framework, the quantity, quality, and total functionality of the investigated hospital is shown in Figure 7-4. However, to investigate the effect of the availability of the basic events on the hospital functionality,

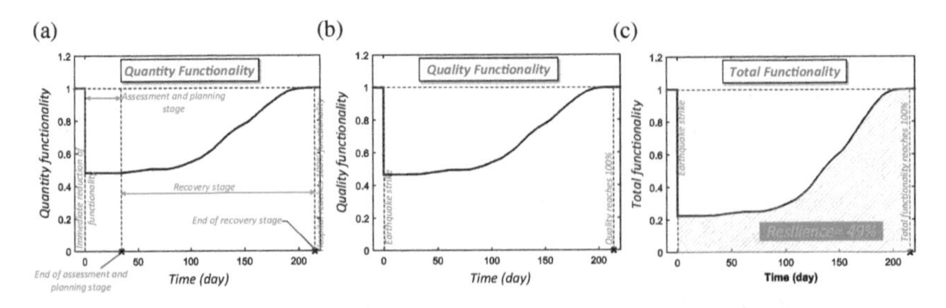

Figure 7-4. Components of the investigated hospital's functionality: (a) quantity functionality, (b) quality functionality, (c) total functionality and resilience.

different values are assigned for these basic events. Five different lifelines, in addition to the hospital, are investigated, namely, power, transportation, telecommunication, water, and sewer networks. Damage for these lifelines is calculated according to Hazus MH 2.1 (Hazus-MH 2015), whereas recovery is calculated using a Markov chain stochastic model coupled with dynamic optimization.

The hospital functionality framework is utilized to estimate the effect of mutual aid, working additional hours, and the availability of alternate staff on the functionality and resilience of the hospital. It also considers the possibility of a staff shortage that results from staff injuries during an earthquake, staff difficulties to go to work, and staff dislocation. Figure 7-5 displays the effect of receiving additional staff, immediately after an earthquake occurrence, on the hospital staff availability. For comparison, an analysis is conducted where it is assumed that the investigated hospital is not provided with alternate staff and does not hire more staff during the recovery time (i.e., 0% additional staff). It can be noted from the figure that receiving additional staff will increase the hospital staff availability, as expected, but it can only enhance the hospital resilience to a certain extent if hospital staff availability is less than space and supplies as shown in Figure 7-5(b). Assuming that the hospital lost 25% of its staff during an earthquake, then staff availability will be the main factor controlling hospital functionality and the effect of the additional staff on the hospital's resilience will be significant as shown in Figure 7-5(c).

To highlight the effect of backup systems, different functionality levels are assigned to the hospital backup systems—100% functional, 10% functional, and 0% functional. In the 100% functional backup systems case, it is assumed that the hospital's backup systems did not suffer any damage during an earthquake. The case for the 0% functional backup systems represents total damage to the backup systems during an earthquake. The 10% functionality level for the backup systems represents severe damage during an earthquake. Figure 7-6 shows how the presence of backup

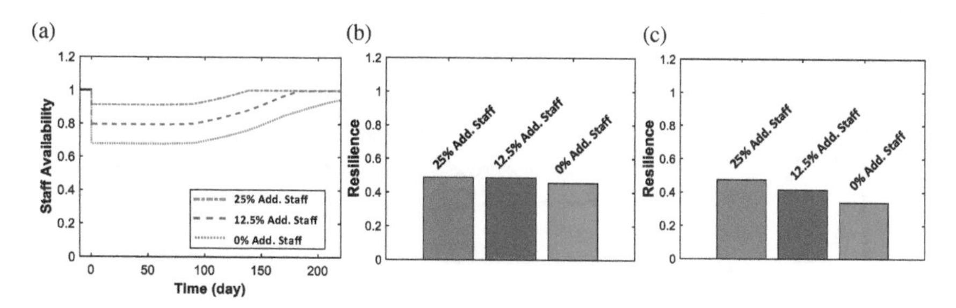

Figure 7-5. The effect of receiving additional staff on (a) hospital's staff availability, (b) hospital's resilience, (c) hospital's resilience for the hospital with 75% initial staff availability.

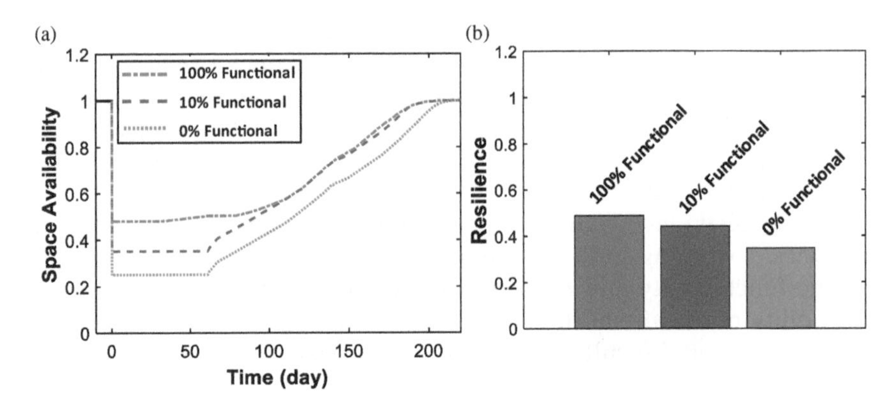

Figure 7-6. The effect of backup systems on (a) hospital's space availability, (b) hospital's resilience.

systems, even if they had experienced significant damage, can keep the hospital functional and resilient after major earthquakes. A hospital without backup systems is expected to sustain lower functionality levels, especially immediately after an earthquake occurrence.

To highlight the importance of having backup space after an earthquake, a different number of extra beds are added, immediately after an earthquake occurrence, which includes 20 and 50 backup beds. The case of 0 backup beds is also investigated. Figure 7-7 shows the effect of providing backup beds on working space availability for the investigated hospital. However, in this situation, increasing the functionality of the working space will have no impact on hospital resilience because the reduction of the supporting lifelines and shortage of medical supplies are the most significant factors for the drop in hospital functionality. Assuming that the hospital had severe content losses that exceeded 75% during an earthquake so that content availability controls the hospital functionality, then the effect of the backup beds on the hospital's resilience will be significant, as shown in Figure 7-7(c).

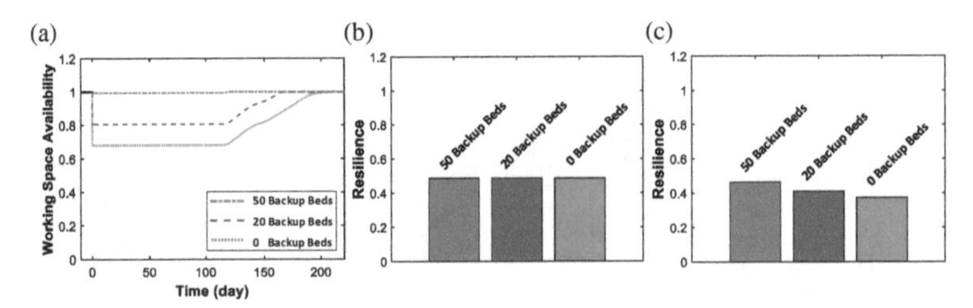

Figure 7-7. The effect of backup space on (a) hospital's working space availability, (b) hospital's resilience, (c) hospital's resilience for the hospital with 25% initial content availability.

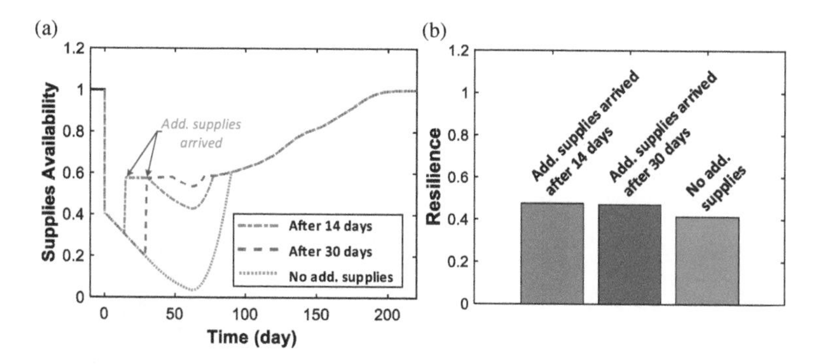

Figure 7-8. The effect of additional supplies on (a) hospital's supplies' availability, (b) hospital's resilience.

To indicate the importance of supply availability, different scenarios are assumed, which include no additional supplies, one additional shipment of supplies arriving 30 days after an earthquake, and one additional shipment of supplies arriving 14 days after an earthquake. Assuming that the amount of supplies in the hospital before the earthquake is only 60% and the average percentage of damage in the supplies' storage rooms is 68%, the percentage of supply availability immediately after the earthquake will be around 41%. Figure 7-8 displays the impact of the supply shortage on the investigated hospital's functionality and resilience as well as the effect of the arrival time of the additional supplies. It is recommended to send additional supplies immediately after earthquakes to the impacted hospital to maintain the highest possible functionality. Providing hospitals with sufficient supplies especially after earthquakes will also limit the chance of the spread of infectious and noninfectious diseases (WHO 2010).

It can be concluded from the previous discussion that achieving an acceptable hospital functionality level after an earthquake is not a simple task and providing enhancement for some component, to reflect changes in a single basic event, might not lead to an increase in hospital resilience. Therefore, investigating and identifying the main reasons for the reduction in hospital functionality are essential steps that can have a substantial impact on the availability of the healthcare service for the investigated community. In this section, three different cases are applied to the investigated hospital. The first case (Scenario_1) can be considered as the best case that comprises all previously mentioned enhancements for the hospital, including 25% additional staff, 100% functional backup system, 50 additional beds, and additional supplies. The second case (Scenario_2) is the existing case. The third case (Scenario_3) is the worst case in which the hospital is not provided with additional staff, backup systems, backup space, or additional supplies. Figure 7-9 shows the effect of different scenarios on the hospital's total functionality and resilience.

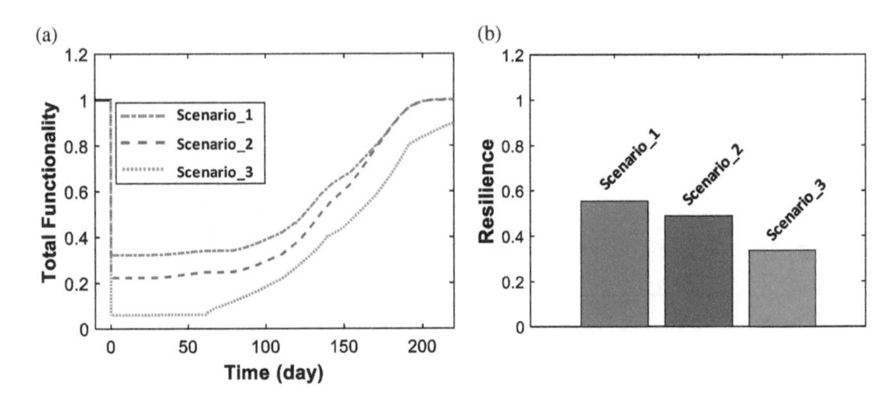

Figure 7-9. The effect of different scenarios on (a) hospital's total functionality, (b) hospital's resilience.

The hospital resilience significantly increases in Scenario_1 because the functionality is only controlled by damage to the hospital itself, which increases hospital independence as recommended by FEMA 577 (FEMA 2007). In Scenario_1, the initial drop of functionality is also reduced, which increases both the quality and quantity of the investigated hospital and increases the overall functionality of the healthcare service.

7.5 RECOMMENDED PRACTICES

In this chapter, the factors affecting healthcare functionality and resilience after earthquakes are discussed. Lessons from previous events and guidelines are merged with the functionality and resilience estimations that resulted from an analytical framework that applied to a midsize hospital located in Memphis, Tennessee. An earthquake with an intensity of 1.0 g is assumed to have struck the hospital and its supporting lifelines. The hospital recovery is estimated based on different scenarios. The following main conclusions can be drawn from the chapter:

1. Ensuring the existence of alternate staff, offering regular training for the staff to increase their preparedness for disasters, and establishing mutual-aid agreements with other hospitals are key to hospital operation following natural hazard events.
2. Maintaining the availability of utility backup systems as well as backup spaces is essential and can significantly impact hospitals' functionality.
3. Securing different providers for the main services that hospitals require and relying on multiple suppliers are pivotal to functionality.

4. Reducing hospitals' supplies after earthquakes can lead to catastrophic consequences. Therefore, receiving the required supplies on time is vital for maintaining an acceptable level of functionality.
5. Organizing healthcare services between hospitals and other healthcare facilities especially after earthquakes is the fundamental component to ensure that most patients will receive the appropriate service, reducing mortality rates as a result.
6. Distributing repair resources among healthcare facilities and other infrastructure requires careful investigation to ensure that there is a balance between the social and economic stability of communities.

REFERENCES

Achour, N., M. Miyajima, M. Kitaura, and A. Price. 2011. "Earthquake-induced structural and nonstructural damage in hospitals." *Earthquake Spectra* 27 (3): 617–634.

Achour, N., F. Pascale, A. D. F. Price, F. Polverino, K. Aciksari, M. Miyajima, et al. 2016. "Learning lessons from the 2011 Van earthquake to enhance healthcare surge capacity in Turkey." *Environ. Haz.* 15 (1): 74–94.

Alcalde-Castro, J., T. Hernández-Gilsoul, I. Domínguez-Rosado, Y. Chavarri-Guerra, and E. Soto-Perez-de-Celis. 2018. "Cancer care after the 2017 central Mexico earthquake." *J. Clin. Oncol.* 4: 1–4.

Almufti, I., and M. Willford. 2013. *REDiTM rating system: Resilience-based earthquake design initiative for the next generation of buildings*. London: Arup.

Arboleda, C. A. 2006. *Vulnerability assessment of the operation of health care facilities during disaster events*. West Lafayette, IN: Purdue University.

Arboleda, C. A., D. M. Abraham, and R. Lubitz. 2007. "Simulation as a tool to assess the vulnerability of the operation of a health care facility." *J. Perform. Constr. Facil.* 21 (4): 302–312.

ASCE. 2010. *Minimum design loads for buildings and other structures*. ASCE 7–10. Reston, VA: ASCE.

Asplin, B. R., T. J. Flottemesch, and B. D. Gordon. 2006. "Developing models for patient flow and daily surge capacity research." *Acad. Emergency Med.* 13 (11): 1109–1013.

ATC (Applied Technology Council). 1985. *Earthquake damage evaluation data for California*. ATC-13. Redwood City, CA: ATC.

Berche, B., C. Von Ferber, T. Holovatch, and Y. Holovatch. 2009. "Resilience of public transport networks against attacks." *Eur. Phys. J. B* 71 (1): 125–137.

Bilgin, H. 2016. "Generation of fragility curves for typical RC health care facilities: Emphasis on hospitals in Turkey." *J. Perform. Constr. Facil.* 30 (3): 04015056.

Burton, H. V., G. Deierlein, D. Lallemant, and T. Lin. 2016. "Framework for incorporating probabilistic building performance in the assessment of community seismic resilience." *J. Struct. Eng.* 142 (8): C4015007.

Ceferino, L., J. Mitrani-Reiser, A. Kiremidjian, and G. Deierlein. 2020. "Effective plans for hospital system response to earthquake emergencies." *Nat. Commun.* 11 (1): 1–12.

Chand, A. M., and M. Loosemore. 2016. "Hospital disaster management's understanding of built environment impacts on healthcare services during extreme weather events." *Eng. Constr. Archit. Manage.* 23 (3): 385–402.

Chen, X., J. Li, and J. Cheang. 2010. "Seismic performance analysis of Wenchuan hospital structure with viscous dampers." *Struct. Des. Tall Spec. Build.* 19 (4): 397–419.

Cimellaro, G. P. 2016. *Urban resilience for emergency response and recovery.* Cham, Switzerland: Springer.

Cimellaro, G. P., V. Arcidiacono, A. M. Reinhorn, and M. Bruneau. 2013. "Disaster resilience of hospitals considering emergency ambulance services." In *Structures Congress 2013: Bridging Your Passion with Your Profession*, edited by B. J. Leshko and J. McHugh, 2824–2836. Reston, VA: ASCE.

Cimellaro, G. P., and M. Pique. 2016. "Resilience of a hospital emergency department under seismic event." *Adv. Struct. Eng.* 19 (5): 825–836.

Cimellaro G., A. Reinhorn, and M. Bruneau. 2005. "Seismic resilience of a health care facility." In *Proc., Ann. Meeting Asian Pacific Network of Centers for Earthquake Engineering Research (ANCER)*, November 10–13, Seogwipo KAL Hotel, Jeju, Korea, Paper N3-Session III.

Cimellaro, G. P., A. Reinhorn, and M. Bruneau. 2008. *Quantification of seismic resilience of health care facilities*. Tech. Rep. MCEER-09-0009. Buffalo, NY: State University of New York.

Cosenza, E., S. L. Di, G. Maddaloni, G. Magliulo, C. Petrone, and A. Prota. 2015. "Shake table tests for the seismic fragility evaluation of hospital rooms." *Earthq. Eng. Struct. Dyn.* 44 (1): 23–40.

Denver Health. 2005. *National hospital available beds for emergencies and disasters (HAvBED) system.* Final Rep. Rockville, MD: Agency for Healthcare Research and Quality.

Didier, M., M. Broccardo, S. Esposito, and B. Stojadinovic. 2018. "A compositional demand/supply framework to quantify the resilience of civil infrastructure systems (Re-CoDeS)." *Sust. Resil. Infrastruct.* 3(2): 86–102.

Di Sarno, L., G. Magliulo, D. D. Angela, and E. Cosenza. 2018. "Experimental assessment of the seismic performance of hospital cabinets using shake table testing." *Earthq. Eng. Struct. Dyn.* 48 (1): 103–123.

FEMA. 2007. *Risk management series: Design guide for improving hospital safety in earthquakes, floods, and high winds*. FEMA 577. Washington, DC: FEMA.

FEMA. 2010. *Risk management series: Design guide for improving school safety in earthquakes, floods, and high winds*. FEMA P-424. Washington, DC: FEMA.

Ferraioli, M. 2015. "Case study of seismic performance assessment of irregular RC buildings: Hospital structure of Avezzano (L'Aquila, Italy)." *Earthq. Eng. Eng. Vib.* 14 (1): 141–156.

GOH (Government of the Republic of Haiti). 2010. *Action plan for national recovery and development of Haiti*. Port-Au-Prince: GOH.

Guinet, A., and R. Faccincani. 2016. "Hospital's vulnerability assessment." In *Proc., 2015 Int. Conf. on Industrial Engineering and Systems Management*, 249–254. Piscataway, NJ: IEEE.

Hassan, E. M., and H. Mahmoud. 2017. "Modeling resolution effects on the seismic response of a hospital steel building." *J. Constr. Steel Res.* 139: 254–271.

Hassan, E. M., and H. Mahmoud. 2018. "A framework for estimating immediate interdependent functionality reduction of a steel hospital following a seismic event." *Eng. Struct.* 168: 669–683.

Hassan, E. M., and H. Mahmoud. 2019. "Full functionality and recovery assessment framework for a hospital subjected to a scenario earthquake event." *Eng. Struct.* 188: 165–177.

Hassan, E. M., and H. Mahmoud. 2020. "An integrated socio-technical approach for post-earthquake recovery of interdependent healthcare system." *Reliab. Eng. Syst. Saf.* 201: 106953.

Hazus-MH. 2015. *Multi-hazard loss estimation methodology: Earthquake model*. Washington, DC: FEMA.

Heaslip, K., W. Louisell, J. Collura, and N. Urena Serulle. 2010. "A sketch level method for assessing transportation network resiliency to natural disasters and man-made events." In *Proc., Transportation Research Board 89th Annual Meeting*, 1–15. Washington, DC: Transportation Research Board.

Hiete, M., M. Merz, C. Trinks, and F. Schultmann. 2011. "Scenario-based impact analysis of a power outage on healthcare facilities in Germany." *Int. J. Disaster Resilience Built Environ.* 2 (3): 222–244.

IASC (Inter-Agency Standing Committee). 2005. *Pakistan earthquake October 2005 consolidated health situation. Bull. No. 2.* Geneva: IASC.

IOM (Institute of Medicine). 2015. *Healthy, resilient, and sustainable communities after disasters: Strategies, opportunities, and planning for recovery*. Washington, DC: National Academies Press.

Isumi, M., N. Nomura, and T. Shibuya. 1985. "Simulation of post-earthquake restoration for lifeline systems." *Int. J. Mass Emerg. Disast.* 3 (1): 88–105.

Jacques, C. C., J. McIntosh, S. Giovinazzi, T. D. Kirsch, T. Wilson, and J. Mitrani-Reiser. 2014. "Resilience of the Canterbury hospital system to the 2011 Christchurch earthquake." *Earthquake Spectra* 30 (1): 533–554.

Jennings, P. C., and G. W. Housner. 1971. *The San Fernando, California, Earthquake of February 9, 1971*. Washington, DC: US Geological Survey and National Oceanic and Atmospheric Administration.

Karapetrou, S., M. Manakou, D. Bindi, B. Petrovic, and K. Pitilakis. 2016. "Time-building specific" seismic vulnerability assessment of a hospital RC building using field monitoring data." *Eng. Struct*. 112: 114–132.

Kirsch, T. D., J. Mitrani-Reiser, R. A. Bissell, L. M. Sauer, M. Mahoney, W. T. Holmes, et al. 2010. "Impact on hospital functions following the 2010 Chilean earthquake." *Disast. Med. Public Health Prepared*. 4 (2): 122–128.

Korkmaz, K. A., A. Nuhoglu, B. Arisoy, and A. I. Carhoglu. 2012. "Investigation of structural safety of existing masonry healthcare facilities in Turkey." *J. Vib. Cont*. 18 (6): 867–877.

Kozin, F., and H. Zhou. 1990. "System study of urban response and reconstruction due to earthquake." *J. Eng. Mech*. 116 (9): 1959–1972.

Kulshrestha, A., and J. Singh. 2016. "Inter-hospital and intra-hospital patient transfer: Recent concepts." *Indian J. Anaesth*. 60 (7): 451–457.

Levinson, D. R. 2014. *Hospital emergency preparedness and response during Superstorm Sandy*. Washington, DC: Dept. of Health and Human Services, Office of the Inspector General.

Lin, P., and N. Wang. 2017. "Stochastic post-disaster functionality recovery of community building portfolios I: Modeling." *Struct. Saf*. 69: 96–105.

Lupoi, A., F. Cavalieri, and P. Franchin. 2013. "Seismic resilience of regional health-care systems." In *Proc., 11th Int. Conf. Struct. Saf. Reliab. Safety, Reliability, Risk and Life-Cycle Performance of Structures and Infrastructures*, edited by G. Deodatis, B. R. Ellingwood, and D. M. Frangopol, 4221–4228. London: Taylor & Francis.

Mahmoud, H., and A. Chulahwat. 2018. "Spatial and temporal quantification of community resilience: Gotham city under attack." *Comput. Aided Civ. Infrastruct. Eng*. 33 (5): 353–372.

Malavisi, M., G. P. Cimellaro, V. Terzic, and S. Mahin. 2015. "Hospital emergency response network for mass casualty incidents." In *Proc., Structures Cong. 2015*, edited by N. Ingraffea and M. Libby, 1573–1584. Reston, VA: ASCE.

Masko, M. L., C. M. Eckert, N. H. M. Caldwell, and P. J. Clarkson. 2011. "Designing for resilience: Using a Delphi study to identify resilience issues for hospital designs in a changing climate." In Vol. 5 of *Proc., ICED 11–18th Int. Conf. on Engineering Design—Impacting Society through Engineering Design*, edited by S. J. Culley, B. J. Hicks, T. C. McAloone, T. J. Howard, and J. Malmqvist, 60–69, Denmark: Design for X/Design to X, Lyngby/Copenhagen.

Maxwell, J. R. 1984. "Quality assessment in health." *BMJ* 288 (6428): 1470–1472.

McCarthy, K., H. M. Mcgee, and C. A. O. Boyle. 2000. "Outpatient clinic waiting times and non-attendance as indicators of quality." *Psychol. Health Med*. 5 (3): 287–293.

McCarthy, M. L., D. Aronsky, and G. D. Kelen. 2006. "The measurement of daily surge and its relevance to disaster preparedness." *Acad. Emergency Med.* 13 (11): 1138–1141.

McDaniels, T., S. Chang, D. Cole, J. Mikawoz, and H. Longstaff. 2008. "Fostering resilience to extreme events within infrastructure systems: Characterizing decision contexts for mitigation and adaptation." *Global Environ. Change* 18 (2): 310–318.

McIntosh, J. K., C. Jacques, J. Mitrani-Reiser, T. D. Kirsch, S. Giovinazzi, and T. M. Wilson. 2012. "The impact of the 22nd February 2011 earthquake on Christchurch Hospital." In *Proc., 2012 NZSEE Conf.*, 1–11. Wellington, New Zealand: New Zealand Society for Earthquake Engineering.

Miniati, R., and C. Iasio. 2012. "Methodology for rapid seismic risk assessment of health structures: Case study of the hospital system in Florence, Italy." *Int. J. Disaster Risk Reduct.* 2: 16–24.

Mulyasari, F., S. Inoue, S. Prashar, K. Isayama, M. Basu, N. Srivastava, et al. 2013. "Disaster preparedness: Looking through the lens of hospitals in Japan." *Int. J. Disaster Risk Sci.* 4 (2): 89–100.

NEHRP Consultants Joint Venture. 2013. *Cost analyses and benefit studies for construction in Memphis, Tennessee (design drawings)*. Gaithersburg, MD: NEHRP Consultants Joint Venture.

Nielsen, M. J., S. Ferguson, A. K. Joshi, S. Rimal, I. Shrestha, and S. R. Magar. 2016. "Post-earthquake recovery in Nepal." *Lancet Global Health* 4: e161.

Nikfar, F., and D. Konstantinidis. 2017. "Shake table investigation on the seismic performance of hospital equipment supported on wheels/casters." *Earthq. Eng. Struct. Dyn.* 46 (2): 243–266.

Nikfar, F., and D. Konstantinidis. 2019. "Experimental study on the seismic response of equipment on wheels and casters in base-isolated hospitals." *J. Struct. Eng.* 145 (3): 04019001.

NIST (National Institute of Standards and Technology). 2016. Vol. 2 of *Community resilience planning guide for buildings and infrastructure systems*. Gaithersburg, MD: NIST.

Ochi, S., M. Tsubokura, S. Kato, S. Iwamoto, S. Ogata, T. Morita, et al. 2016. "Hospital staff shortage after the 2011 triple disaster in Fukushima, Japan-an earthquake, tsunamis, and nuclear power plant accident: A Case of the Soso district." *PLoS One* 11 (10): e0164952.

Paterson, J., P. Berry, K. Ebi, and L. Varangu. 2014. "Health care facilities resilient to climate change impacts." *Int. J. Environ. Res. Public Health* 11 (12): 13097–13116.

Peek-Asa, C., J. F. Kraus, L. B. Bourque, D. Vimalachandra, J. Yu, and J. Abrams. 1998. "Fatal and hospitalized injuries resulting from the 1994 Northridge earthquake." *Int. J. Epidemiol.* 27 (3): 459–465.

Perrone, D., M. A. Aiello, M. Pecce, and F. Rossi. 2015. "Rapid visual screening for seismic evaluation of RC hospital buildings." *Structures* 3: 57–70.

Petrazzi, L., R. Striuli, L. Polidoro, M. Petrarca, R. Scipioni, M. Struglia, et al. 2013. "Causes of hospitalisation before and after the 2009 L'Aquila earthquake." *Inter. Med. J.* 43 (9): 1031–1034.

Ripp, J. A., J. Bork, H. Koncicki, and R. Asgary. 2012. "The response of academic medical centers to the 2010 Haiti earthquake: The Mount Sinai school of medicine experience." *Am. J. Trop. Med. Hyg.* 86 (1): 32–35.

Rodgers, J., V. Cedillos, H. Kumar, L. T. Tobin, and K. Yawitz. 2009. *Reducing earthquake risk in hospitals from equipment, contents, architectural elements and building utility systems.* New Delhi, India: Geohazards International and Geohazards Society.

Schultz, C. H., K. L. Koenig, and R. J. Lewis. 2003. "Implications of hospital evacuation after the Northridge, California, Earthquake." *N. Engl. J. Med.* 348 (14): 1349–1355.

Shang, Q., T. Wang, and L. Jichao. 2020. "Seismic resilience assessment of emergency departments based on the state tree method." *Struct. Saf.* 85: 101944.

Sprivulis, P. C., J. Da Silva, I. G. Jacobs, G. A. Jelinek, and A. R. L. Frazer. 2006. "The association between hospital overcrowding and mortality among patients admitted via Western Australian emergency departments." *Med. J. Aust.* 184 (5): 208–212.

Stratton, S. J., and R. D. Tyler. 2006. "Characteristics of medical surge capacity demand for sudden-impact disasters." *Acad. Emergency Med.* 13 (11): 1193–1197.

TariVerdi, M., E. Miller-Hooks, T. Kirsch, and S. Levin. 2019. "A resource-constrained, multi-unit hospital model for operational strategies evaluation under routine and surge demand scenarios." *IISE Trans. Healthcare Syst. Eng.* 9 (2): 103–119.

Tekin, E., A. Bayramoglu, M. Uzkeser, and Z. Cakir. 2017. "Evacuation of hospitals during disaster, establishment of a field hospital, and communication." *Eurasian J. Med.* 49 (2): 137–141.

Therrien, M. C., J. M. Normandin, and J. L. Denis. 2017. "Bridging complexity theory and resilience to develop surge capacity in health systems." *J. Health Organ. Manage.* 31 (1): 96–109.

UNICEF (United Nations Children's Fund). 2004. *Crisis appeal earthquake in Bam, Iran.* New York: UNICEF.

Villegas, P., E. Malkin, and K. Semple. 2017. "Mexico earthquake, strongest in a century, kills dozens." *New York Times.*

Vugrin, E. D., S. J. Verzi, P. D. Finley, M. A. Turnquist, A. R. Griffin, K. A. Ricci, et al. 2015. "Modeling hospitals' adaptive capacity during a loss of infrastructure services." *J. Healthcare Eng.* 6 (1): 85–120.

Wang, Y. 2014. *Hospital and water system earthquake risk evaluation.* Salem, OR: Oregon Health Authority.

Watson, S. K., J. W. Rudge, and R. Coker. 2013. "Health systems' "surge capacity": State of the art and priorities for future research." *Milbank Q.* 91 (1): 78–122.

WHO (World Health Organization). 2003. *Guidelines on the offer and acceptance of field hospitals for use in Iraq*. Washington, DC: WHO.

WHO. 2010. *Public health risk assessment and interventions: Earthquake Haiti.* Geneva: WHO.

Youance, S., and M. Nollet. 2012. "Post-earthquake functionality of critical facilities: A hospital case study." In *Proc., 15th World Conf. on Earthquake Engineering*, Lisbon, 1–10. Tokyo: International Association of Earthquake Engineering.

CHAPTER 8
RESILIENCE AND CONSTRUCTION

Hamid R. Adib

8.1 INTRODUCTION

The traditional definition of resilience focuses on the ability to absorb or avoid damage without suffering complete failure. However, this definition is rather narrow, and, therefore, the important topic of resilience deserves a more comprehensive explanation. A more appropriate definition of resilience, especially as it pertains to the built environment, encompasses the four principal areas of robustness, resourcefulness, recovery, and redundancy.

As engineers and construction professionals, we simultaneously focus on optimization and failure avoidance. We learn from past mistakes and develop feedback systems to improve design and construction processes, and as part of this learning, we strive to challenge the norm and push the limits. As part of this process, one should have a disaster recovery and continuity of business plan. Disaster recovery, by assuming that an unusual event has occurred, is the action of bringing the process of delivery of the project that was affected by the event to the before-the-incident condition (Nigg 1995). Business continuity is the risk or threat management plan that allows the business to continue smoothly against disruptions to operations. Disaster recovery and business continuity are intertwined and consider scenarios for disruptions and appropriate preventative or mitigation measures as part of a comprehensive resilience plan.

Resilient construction planning also requires an optimized solution that lies in embracing the notion that resilience is a multidisciplinary subject and not limited to a field of practice. It also requires one to consider natural hazards, man-made hazards, and other plausible unusual events. The advantage that the construction community enjoys is resourcefulness stemming from the tools and equipment frequently used at a construction site.

8.2 RESILIENCE CONSIDERATIONS IN THE DESIGN AND CONSTRUCTION COMMUNITY

Traditional planning processes focus on two broad components: anticipation of consequences of an event and proper response to absorb or mitigate the disruption caused by such an event at every step. Studies have shown that without proper planning, the recovery costs can be astronomical, and hence, proper attention to resilient design and construction is warranted.

As a professional community, we reach deeper and recognize the resilience and interdependence of its key features—robustness, resourcefulness, recovery, and redundancy in planning, design and the execution of successful projects. On the broader level of real community, the Federal Emergency Management Agency (FEMA), through its 2018 to 2022 strategic plan (FEMA 2021), embraces the holistic approach to preparedness in collaboration with the state and local governments and the design and construction community—a collaborative top-down bottom-up approach.

Examples of top-down leadership in resilient design are plenty. Following the aftermath of the Superstorm Sandy, New York State created the Governor's Office of Storm Recovery (GOSR n.d.) to support and fund rebuilding projects with resilience as the main focus of reconstruction. New York City is an example of a city that created its own special initiative for inclusion of resiliency in rebuilding through a newly created Office of Recovery and Resiliency. Multiple large projects in the states of New York, New Jersey, and Connecticut are in various stages of design and implementation under these or similar programs.

On the private level, developers, especially those in coastal cities, have made the promotion of resilience strategies a cornerstone of their future development projects and have mobilized resources through a global access to a network of experts. Professional organizations such as the American Institute of Architects (AIA) advocates for the architect's role in resiliency planning and direct assistance to communities through a network of experts to help before, during, and after an event. ASCE, through its Engineering Mechanics Institute, has dedicated resources to study resilience objectively and to address resilience in collaboration with governmental agencies. The engineering community and the American Council of Engineering Companies have embraced resilience design and continuously promote education in resilience.

Engineers have also adopted new paradigms in design. As an example, on several recent projects, Jacobs has been involved in the evaluation of the economics of location of sensitive mechanical and electrical equipment. Whereas traditionally, mechanical, and electrical equipment were installed in basements and subbasements, we explore ideas with the owners and evaluate cost–benefit scenarios to place such critical equipment on higher floors investing in potentially additional ductwork piping and cable routing. Such costs when compared with the costs of losing services or business continuity losses often prove nonconsequential.

The Resilient Design Institute has a list of recommended design strategies such as locating critical elements within a building to withstand severe events and introducing redundancy in utility supplies among many others. The US Green Building Council (USGBC n.d.) has announced that it has adopted the RELi Resilience Rating System with almost 200 different metrics and indicators. The RELi rating system focuses on the specific issues related to resilient design and for projects at all scales of development, from individual buildings to entire communities.

8.3 THE 4RS

The terms *sustainability, resilience,* and *risk* are frequently used interchangeably; however, it is important to emphasize that while there are similarities, they have distinct differences. Risk defines the relationship between a hazard that negatively impacts an asset or a process and the consequences of the event taking place. In simple terms, risk is a function of consequence threat and vulnerability, whereas resilience is related to robustness, resourcefulness, recovery, and redundancy (4Rs). It has been shown that sustainability can be considered a subset of resilience and resilience and the 4Rs a subset of consequence, threat, and vulnerability, and, therefore, risk (Ettouney and Alampalli 2012).

Although the essential components of resilience have been defined in other chapters of this manual, it is worth a refresh while we focus on their applicability to construction and project execution. *Robustness* refers to comprehensive planning and building the capability within the system to maintain critical operations and function, which puts focus on the holistic approach to planning for postdisaster functionality identification and reinforcement of weak links in the system. *Resourcefulness* indicates planning for preparation and the ability to respond to disruption of operations. *Recovery*, more important, rapid recovery, is the preparation and plans put in place for postdisaster ability to return to functionality and operations quickly. *Redundancy* suggests the backup resources that can quickly be tapped into in case of failure.

In this chapter, we discuss resilience in the field of building and infrastructure construction as a thought stimulator.

8.4 ROBUSTNESS

Improving robustness helps resist shocks to the system and ensures efficient operations. Proper planning analysis and management of potential events are the critical elements of an economically viable and robust solution. FEMA, as part of its website, has a location-specific rich database of past natural hazards events (FEMA n.d.) that can help planners obtain the type and magnitude of events that a project can potentially be exposed to based on historical events. "Project Impact: Building Disaster-Resistant Communities," an initiative that was put in place in 1997 following Hurricane Floyd, is a good example of how FEMA is working with communities and government leaders to raise awareness and preparedness. The aim of the initiative is to look beyond financial comfort that flood insurance or government assistance provides. The government documentation and evaluation requirements can be lengthy and the disruption to the community post event is often cumbersome. Building robustness into the system will better prepare communities for events and to be less reliant on the government's speed and level of response. FEMA's 2018 to 2022 Strategic Plan set its number one goal as "building a culture of preparedness." Although FEMA is charged to push the agenda of resilience forward, other federal, state, local, and nongovernmental organizations play a vital role in bringing the diversity of thought and local overall knowledge to the effort.

Building robustness into a *project* is necessary; however, it is not adequate to properly address the topic of resilience of a *system* (community). Catastrophic events are usually widespread, affecting systems or communities with complex interdependencies. Properly planned and designed systems require a recognition of, and attention to, the interdependencies to be better prepared and a proper functioning of critical operations. A systemwide robustness plan requires detailed attention to the vulnerable links in an ecosystem and its overall performance. An example is to address the need to provide critical access to those trapped in remote areas.

Ready access to proper equipment and first responders' knowledge of the availability of equipment and vehicles capable of traveling through rough terrain would be critical. Local heavy civil constructors usually have the required equipment and play an important role to maintain critical access and assist in restoring critical operations and functions. It is equally important to ensure that the equipment is in a work-ready condition and spare parts are available.

8.5 RESOURCEFULNESS

Resourcefulness is essential and the resources can be tools as simple as readily available mobile generators or ensuring availability of spare parts for the equipment, especially those parts that can easily be damaged in case of an event. Supply chain management is an important topic to consider in planning for resources.

The construction supplier industry is populated by small businesses with an extremely complex network of supply chain. On larger projects, the number of separate supplying businesses runs into hundreds if not thousands. Proper resource planning needs to consider effective management and relationship of the entire supply chain to achieve best value, while ensuring robustness of the supply chain. The COVID-19 pandemic greatly tested the previous supply chain system. Plenty of examples exist where hospitals and healthcare organizations paid huge premiums (factors of over 10 times regular pricing) on their normal supplies. However, most organizations have now revamped their supply chain plan and are getting their resources in closer proximity of their operations even if normal costs are marginally higher than before.

Other examples of resourcefulness are to adopt measures such as a more expensive yet reliable satellite communication system. While the suggestion of a satellite communication system may seem farfetched in an urban environment, it has proven useful. Following the catastrophic events of September 11, 2001, many cell phones in New York City did not operate, because major cell phone antennas were located on the collapsed twin towers of the World Trade Center. Priority to restore cell phone service was given to those with special need and first responders who were already registered with the mobile carriers. Those with satellite-operated cells had an advantage of being able to communicate.

Resourcefulness also applies to the availability and readiness of expert manpower. Investigations into the British Petroleum (BP) oil-drill platform's explosion (Gröndahl et al. 2010) revealed a late response because of a lack of resources and available expert manpower as a major contributor to the event quickly become a crisis.

8.6 RECOVERY

Recovery, more important, postdisaster rapid recovery, refers to preparations and plans put in place to enable the users to return to operations quickly. However, quick recovery is not purely a function of business operations and might need to deal with higher urgency needs. An example is the events of August 2010 in Chile where 33 miners were trapped during mining operations in a collapsed mine at a depth of approximately 700 m. Fortunately, all the miners were rescued, although it took 17 days to determine the condition of the miners. Postevent investigations of the preparedness plans revealed that a disorganized safety protocol and a lack of emergency plans were partly responsible for the length of time for the occurrence of the event and the recovery operations to take place (Bonnefoy 2013). The rescue efforts utilized innovative techniques and were fortunately successful, however, at higher costs as compared with the scenario of following restricted safety protocols and proper planning and preparedness. Rapid recovery requires emergency preparedness and a trained team of first responders, but equally important, a detailed safety protocol, well-planned and practiced procedures, and easy access to proper equipment and to spare parts.

The Project Management Institute's publication PMBOK—Project Management Body of Knowledge (PMI 2019)—published by the American National Standard Institute offers a rational focus to developing a model for efficient management of disaster recovery operations. It is recommended that a disaster management board be identified and be at the center of the process with the major task of driving planning and implementation. Pictorial planning of the recover process using the rich picture diagram (RPD) has proven helpful. Examples of the RPD are simple hand sketches or complex relationship flowcharts that can be generated using low-cost collaboration software. Focus areas for the model should, as a minimum, include the following: communications—to address the communication protocol and technology human resources and claims management to manage behaviors during and after an event and a recovery plan for business recovery and continuity protocol that aligns with financial aspects and cost management.

8.7 REDUNDANCY

Most often, planners focus on designing or accommodating utility generation systems (power-chilled water and hot water). In practice or in the literature, terms such as $N + 1$ are used to signify the level of additional resources necessary to achieve a redundant plan (Woo and Grayson 2018). For example, $N + 1$ redundant utility refers to planning for one level of backup in the loss of the normal utility system or service. A critical system, which can be the lifeline of first responders, usually requires a more robust redundant design. This strategy can deliver higher levels of redundancy often in the form of systems that are paralleled. In this plan, any component in one system can be out of service without affecting the operation of the other systems.

Determining the level of needed redundancy requires a clear definition of the risks involved with the loss of utility including lost production, lost revenue, and possible loss of life. The need for the level and the means of delivering redundancy is usually evaluated in terms of costs and benefits, and, in general, the higher the level of redundancy, the higher the cost, but not always. For example, on planning for power redundancy, one can attempt tackling the challenges involved in this exercise through multiple service sources such as distribution served from separate substations or grid services from separate utilities and substations and should consider the stability and reliability of the entire system.

Renewable or alternate power sources like photovoltaic systems and fuel cells or uninterruptible power supplies can also be excellent sources of supply service. Backup power must also be provided for the supporting cooling and heating systems, fire detection and suppression systems, lighting and control systems, and any other system required for critical facility operations such as security and access control and so on. Mobile combined heat and water plants provide a plausible option for redundant service. Resilience in a mechanical system is more challenging as this system is multilayered, and for the system to reach a certain redundancy, certain components of the system must be resilient.

8.8 PLAN

Risk is inherent in the construction industry because of the very nature of the industry, and as a result, constructors are very familiar with risk management and mitigation. Risk is made up of the uncertainties and vulnerabilities of a project delivery process, which if it occurs, can change the outcome or the achievement of the objectives of a project. The level of risk management is influenced by the project importance, its complexity execution factor, and internal and external factors. Addressing resilience requires the same rigor in planning and documentation of procedures to ensure that robustness, resourcefulness, recovery, and redundancy are fully thought-through and warranted for a given project or the community that houses the project.

The resilience plan shall also give equal consideration to pre- and postdisaster conditions in building robustness and redundancy into the system. Predisaster planning deals with identification, estimation, and preparation to mitigate risks that can cause an event. Postdisaster planning assumes that an event has occurred and deals with activities to help get normal operations back on track efficiently and quickly.

A properly devised resilience plan shall consider all attributes related to hazard identification, risk management, safety planning, and resource allocation. The probability of the occurrence of an event, sometimes called "hazard analysis," is the estimation of a threat and measuring a hazard effect. We can learn quite a bit by studying the well-established process and procedures in the petrochemical industry. Critical industries exposed to high hazard operations have long faced challenging hazards (traditionally natural and more recently man-made) and have used mitigation identification and analysis techniques such as the what–if the fault-tree HAZOP event-tree checklists and so on to tackle these hazards (Sutton 2015). Similar methodologies can be used in construction to build better resilience in the design and execution while being cognizant of costs.

Preparedness is essential to keep the high standards of safety. The industry documents daily events diligently. Such documents are then used in preparedness of well-thought-through planning for potential events and hazards. Anyone working or even visiting a chemical facility is required to be educated of the potential risks of what can go wrong and what to do in case something does go wrong. The moto is "safety is everyone's responsibility." The journal of events and documentation processes provide input to lessons learned to help with root cause analysis and identify and anticipate potential mishaps. Abiding by the safety rules and risk management protocol are enforced vigorously. Not only are risks very actively managed, a crisis management protocol is put in place to help minimize the effects of an undesirable event and for expeditious return to normal operations.

Planning is an important part of construction management. Constructors spend significant effort in developing detailed construction schedules, paying attention to the risks that can adversely affect the achievement of critical milestones. The plan is first developed broadly and then given granularity with finer details. The thought process that goes into regular planning and risk assessment for construction projects can readily be expanded to identify the steps one can take to improve robustness and resilience as a natural extension of regular construction planning of a project.

First, costs in most cases become decision drivers and cause complexities in the matter of justification of expenditures in preparation for a major event. The identification of potential events is critical in planning, as those that are not properly identified, planned for and mitigated, or mismanaged can lead to crisis. An example is the BP oil-drill platform's explosion on April 2010 in the Gulf of Mexico. The explosion caused eleven deaths and more than seventeen injuries and resulted in extensive pollution to the Gulf of Mexico's waters. Investigations following the event revealed two major miscalculations in building robustness into the execution plan: (1) The response to the emergency and steps to control the situation and limit the implications were not properly planned, and (2) the risks associated with the construction process were not properly mitigated.

A lack of proper resilience planning for events such as the BP explosion in 2010 turns an otherwise regular event into a crisis. Building robustness into the system is a good tool for crisis management.

Unfortunately, in construction projects, there is no generic management theory that simplifies the process of crisis management. Inherent robustness and proper communication tools and protocol seem to be the weak links in crisis management. RPDs based on the grounded theory methodology are proposed as a tool for making a thorough analysis and achieve improved preparedness and communication (Sutrisna and Barrett 2007). The analysis includes an in-depth understanding of potential failures and planning for recovery with an eye on the project management constraints of time and cost, while keeping focus on value and quality. Examples of other considerations for robustness and crisis management are as follows: risk evaluation and safety planning according to the historical data of accidents and disasters, installation of safe work areas, focus on construction method improvement, embracing continuous worker safety training, and effective safety planning.

8.9 ECONOMICS

Economics and costs associated with the plan play an important factor in the process. One will have to consider the costs and benefits of a solution and the methods by which to mitigate the risks of getting the project or the system or the community back on track and properly functioning. While there is still debate on climate change and its impacts, those who live or work in coastal regions, especially those affected by past unusual natural and weather events, understand that investing in resilience is crucial.

Economic considerations require the justification of expenditures, and typically those who have experienced an extreme event will be more amenable to investment with the aim of prevention of another catastrophic outcome. Take for example the superstorm Sandy. New York City was hit with Superstorm Sandy in 2012, and as of 2019, while major milestones have been achieved in bringing New York City to the pre-event operating and functioning city, the efforts to rebuild parts of New York City are ongoing (NYC n.d.). The plan currently in place considers the 4Rs and includes topics such as the following:

- Storm simulations and evacuation,
- Enhancements to long-term forecasting,
- Rapid warning system enhancement,
- Flooding impacts on transportation and infrastructure,
- Flooding impacts on electric power systems,
- Flooding impacts on wastewater infrastructure, and
- Flooding impacts on drinking water systems.

An economically viable approach will consider not only the technical considerations previously outlined but also an assessment of critical facility and the community's economic vulnerabilities, prioritization of community reconstruction, investment strategies along with the resilience metrics for infrastructure investments, and the impacts of strategies on insurance premiums and property market value. Economic motivators such as incentives will help resilience design decisions to proceed to final adoption and implementation. The motivators can be finance-based incentives, technology-based incentives, regulatory-based incentives, and property-market-based incentives.

Financial incentives could encourage the adoption of hazard mitigation by reducing the initial cost of implementation. Financial incentives can be in terms of tax credits and deductible-reduced insurance premiums, public low-interest loan programs, reduced permit fees, fee waivers, and a cost-sharing approach. Regulatory-based incentives include implementing mandatory disclosure of risks, risk assessment in property valuation, and improving building standards, guidelines, and building codes. Technological innovations such as cost-effective retrofitting design solutions promote more advanced ways of achieving risk reduction through the use of new technology innovative solutions and/or advanced analytical techniques.

A focus on raising hazard and vulnerability awareness among stakeholders and an emphasis on the benefits of business continuity planning help bring focus to the economic viability of solutions. The economic and decision-making considerations of risk mitigation to some natural hazards such as earthquakes have been well studied (Egbelakin et al. 2011). A similar approach and framework can be adopted for resilience considerations. The challenge is the complex nature of assessment of the vulnerability of the existing built environment and its interdependencies and interconnections with the wider context of communities.

8.10 ENFORCEMENT

Authorities having jurisdiction on a project (Occupational Safety and Health Administration, Building Officials Fire Marshals, etc.) have enforcement powers to ensure that the health safety and environmental procedure are followed. Currently, their authority does not apply to resilience and preparedness planning, and, therefore, resilience planning becomes voluntary.

Some bigger cities and municipalities, especially those with large coastal exposure, have begun the process of codifying resilience planning. Codes represent the legal minimum enforceable requirements and standards, and it seems reasonable to expect that incorporating sustainability and resiliency into codes will make new and renovated buildings better and more marketable. The International Green Construction Code (ICC 2018) and California's CALGreen building standard (ICC 2019) have published guidelines and standards to do just this.

8.11 CONCLUSIONS

This chapter explored the disaster recovery approach using resilience in the construction framework in parallel with the project management standards based on Project Management Body of Knowledge (PMBOK)'s knowledge areas. We acknowledged the key features and essential components of resilience—robustness, resourcefulness, recovery, and redundancy and their interdependence in a successful planning design and execution of a project. This framework addresses a range of factors from the social, regulatory, institutional, economic, behavioral, and enforcement perspectives related to disaster preparedness. Economic decisions regarding disaster preparedness, risk mitigation, and motivators can be used to encourage voluntary adoption of mitigation, even though regulatory agencies are advancing their efforts to codify the practice. Regardless of the approach, success will depend on the recognition that the framework needs to be adopted as a collaborative effort between stakeholders—public and private entities.

8.12 RECOMMENDATIONS

Past criticism in the area of natural hazard and disaster management has been the lack of a holistic framework. We hope that the approach presented here will address this concern.

In summary, the 4Rs (robustness, resourcefulness, recovery, and redundancy) framework embraces the following general steps. A more specific challenge-oriented approach can be built on top of this framework. The important perspectives of the framework are as follows:

- Risk and safety management planning stage: The processes of risk analysis should begin with the establishment of a disaster management board and its roles and responsibilities in case of an event. The board will then clearly define the desired redundancy levels the robustness to be implemented and consider economic ramifications. The board helps chart the path for resilience planning and assigns a senior officer in charge of the implementation team.
- Emergency response plan: The implementation team defines the emergency response plan and resources allocated for a speedy response to an event. It is critical to practice emergency protocols to ensure seamless execution.
- Preparation planning: The implementation team also defines the scope, steps, time, and controls for the recovery, preparation, building the robustness and redundancy, and allocating resources. The supply

chain in response planning should be clearly identified and the availability of alternate resources checked.

- Communication planning and preparedness stage: The team then addresses the issues of communication protocol and technology, human resources, and claims management to manage behaviors before, during, and after an event.
- Quality planning: Similar to any other plan, regular evaluation and review of the steps of the plan should be checked regularly and adjustments made as necessary. As new technology or equipment becomes available, consider adding such resources to the plan.
- Recovery planning stage: The board and the senior officer address the recovery and continuity protocol which aligns with financial aspects and cost management goals.
- Legal and ethical planning stage: The board also addresses a disaster's implications to human lives and preparation for the rebuild.

REFERENCES

Egbelakin, T., S. Wilkinson, R. Potangaroa, and J. Ingham. 2021. "Enhancing seismic risk mitigation decisions: A motivational approach." *Constr. Manage. Econ.* 29 (10): 1003–1016.

Ettouney, M. M., and S. Alampalli. 2012. *Building resilience*. Washington, DC: National Institute of Building Sciences.

FEMA. 2021. "2018–2022 Strategic plan." Accessed November 2, 2021. https://www.fema.gov/sites/default/files/2020-03/fema-strategic-plan_2018-2022.pdf.

FEMA. n.d. "FEMA database of natural hazards." Accessed November 2, 2021. https://www.fema.gov/disasters/disaster-declarations.

GOSR (Governor's Office of Storm Recovery). n.d. Accessed November 2, 2021. https://stormrecovery.ny.gov/.

ICC (International Code Council). 2018. "The international green construction code." Accessed November 2, 2021. https://www.iccsafe.org/products-and-services/i-codes/2018-i-codes/igcc/.

ICC. 2019. "CalGreen—2019 California Green Building Standards Code Title 24 Part 11." Accessed November 2, 2021. https://calgreenenergyservices.com/wp/wp-content/uploads/2019_california_green_code.pdf.

Gröndahl, M., Park, H., Roberts, G., and Tse, A. 2010. "Investigating the cause of the deepwater horizon blowout." *The New York Times*, June 21, 2010.

Bonnefoy, P. 2013. "Inquiry on mine collapse in Chile ends with no charges." *The New York Times*, August 1, 2013.

Nigg, J. M. 1995. *Wellington after the Quake: The challenge of rebuilding cities*, 81–92. Wellington, New Zealand: Centre for Advanced Engineering and Earthquake Commission.

NYC (New York City). n.d. "New York City recovery." https://www1.nyc.gov/site/recovery/index.page.

PMI (Project Management Institute). 2019. *A guide to the project management body of knowledge (PMBOK® guide)*. ANSI/PMI 99-001-2017. 6th ed. Newton Square, PA: PMI.

Sutrisna, M., and P. Barrett. 2007. "Applying rich picture diagrams to model case studies of construction projects." *Eng. Constr. Archit. Manage.* 14 (2): 164–179.

Sutton, I. 2015. *Process risk and reliability management*. 2nd ed. Houston: Gulf Professional.

USGBC (US Green Building Council). n.d. "Resilient design for a changing world." Accessed November 2, 2021. http://c3livingdesign.org/?page_id=5110stem.

Woo, M., and S. Grayson. 2018. *Resilient utility infrastructure—Part 1: Planning and design*. Wood Harbinger Newsletter Series. Bellevue, WA: Wood Harbinger.

INDEX

Note: Page numbers followed by *f* and *t* indicate figures and tables.